CW00938009

DEER MANAGEMENT IN THE UK

DEER MANAGEMENT IN THE UK

DOMINIC GRIFFITH

visit www.deerengland.co.uk

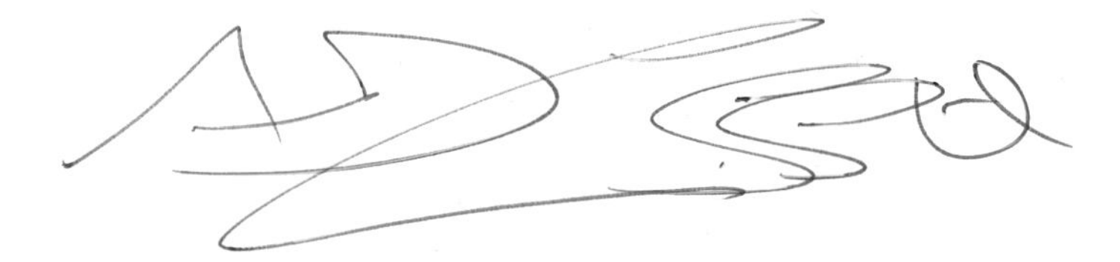

Quiller

First published in the UK in 2004 by the author
This edition published 2011
by Quiller, an imprint of Quiller Publishing Ltd

British Library Cataloguing-in-Publication Data
A catalogue record for this book is available from the British Library

ISBN 978 1 84689 108 3

Photos on pages 1 and 2 by Chris Howard

All other photographs by the author, except where credits shown

Designed and typeset by Paul Saunders

Printed in China

Quiller
An imprint of Quiller Publishing Ltd
Wykey House, Wykey, Shrewsbury, SY4 1JA
Tel: 01939 261616 Fax: 01939 261606
E-mail: info@quillerbooks.com
Website: www.countrybooksdirect.com

CONTENTS

Acknowledgements

I would like to thank the following for their contributions towards the compilation of this book.

The Training Department of the British Deer Society for allowing me to share some of its material.

Richard Prior, for introducing me to the management of roe and to trophy measuring, and for his support in that up until 2007. And to Charles Fenn in particular for his help in updating the records in chapter 12.

Individual members of the UK CIC Trophy Commission for their support over the ten years since its inception in 1997 until 2007.

Shooting Times, for allowing me to make use of material previously published in the magazine.

The landowners who have allowed me the opportunity to demonstrate what it is possible to achieve with deer at estate level.

Fürst zu Löwenstein for his support over almost 20 years, and for so kindly agreeing to write the Foreword.

Marco Pierre White, for generously providing a range of delicious recipes.

Colin Dunton for the photos of Marco Pierre White's perruque.

Chris Howard for allowing me to use some of his exceptional photos of deer from one of the case study areas.

Brian Phipps for the use of his equally wonderful photos.

The many deer enthusiasts who have shared their knowledge and experience with me over many years.

FOREWORD

As a species of game in the UK, the roe deer remains much less understood, appreciated and valued than the pheasant or grouse. Here in Continental Europe we treasure the roe as one of the most fascinating of animals to observe and stalk. They are great individuals, adaptable even to our overpopulated landscapes and above all are responsive to professional management. I am therefore delighted that Dominic Griffith, whom I have come to regard as one of the most learned experts on roe, has brought new material and case studies to this updated edition of his excellent book. We have worked together for over 15 years, learned much from each other and enjoyed the proven results of a joint management philosophy.

This book should help all stalkers and deer managers to better understand these beautiful and captivating animals, and I commend it to all enthusiasts.

ALOIS K FÜRST ZU LÖWENSTEIN

Photo opposite by Chris Howard

INTRODUCTION

It gives me great pleasure to include a few recipes in Dominic Griffith's excellent book. He is without question one of the most knowledgeable stalkers and authoritative deer managers in Great Britain today. We share a passion for the roe and over many years he has taught me a tremendous amount about the management of roe deer and in particular about his low stress management technique. As a measurer he is incredibly thorough and, like Richard Prior and Charles Fenn, shows true understanding of the species. He is also an instinctive stalker, a great teacher and deserves recognition as someone who has actually proved his own theories of management.

MARCO PIERRE WHITE

1.

A MANAGEMENT PHILOSOPHY

DEER MANAGEMENT success can only be measured against a clear idea of the precise aims and objectives of an Estate deer management policy. While a forester, for example, might see success as simply a dramatic reduction in deer numbers, a deer enthusiast might see success in precisely the opposite terms. This book addresses the subject from the standpoint of a land manager who takes the genuinely altruistic view that, as a temporary steward of the land, he wishes simply to manage deer to their best potential for a given environment. Our indigenous species have been here for something like half a million years. Large-scale intensive forestry began less than a hundred years ago with the setting up of the Forestry Commission in 1919. Deer are woodland animals, and I think they have at least an equal right of occupation as other species. Our duty is to manage them according to the best practices available. These are seen as:

- Minimal stress.
- Maximum quality.
- Optimum quantity.

In other words we wish to see as large and healthy a population as the habitat can sustain, but not at the expense of other land management objectives. Deer management is seen as a purposeful and fulfilling task, which can be undertaken compassionately and at minimal stress to the deer, but with the aim of sustaining a predictable population of high

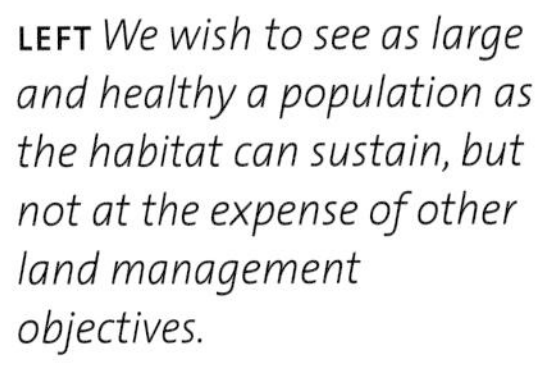

FAR LEFT *Deer are woodland animals, and they have at least an equal right of occupation.*

LEFT *We wish to see as large and healthy a population as the habitat can sustain, but not at the expense of other land management objectives.*

Deer management is seen as a purposeful and fulfilling task, which can be undertaken compassionately and at minimal stress to the deer. Photo by Brian Phipps

quality and at an even sex ratio. Now if this seems rather Utopian, I hope to prove that the dividend of such enlightened policy is a healthy, visible and truly manageable population of wild deer.

The deer management plan that I have espoused and practised over many years is not a personal crusade, but a partnership between myself and a group of enthusiastic landowners to whom I must be ever grateful. One of our clearest successes is visible through the approach we have taken to achieving the cull. By stalking the deer intensively over short periods at a time, we have been able to leave them undisturbed for the greater part of the year. Deer have highly evolved anti-predator reflexes, and our approach to culling must take this into account. Stalked regularly,

A healthy, visible and truly manageable population of wild deer.
Photo by Chris Howard

deer become shy and eventually more nocturnal. Deer shot from motor cars, now sadly legalised by the Regulatory Reform Order (Deer) 2007, become very quickly wary of all human activity and soon become unmanageable. Our deer are therefore relatively easy to approach at all times, and settle again within days or even hours of the brief culling periods, ensuring maximum visibility at all other times. Most census work is done from vehicles, with the deer quickly becoming entirely used to the daily traffic through the forest. Tolerance of vehicles is, in itself, probably the single most important objective to achieve.

If commercialism has formed any part of their management philosophy, then it is only as an occasional by-product of good management, and is intended to do no more than contribute towards the management expenses. Deer management may therefore be able to provide a regular source of revenue to mitigate its associated costs, and in the very best circumstances may even run a small profit. If, however, the primary management aim is to make money, then your aims are almost certainly fatally flawed.

Deer have highly evolved anti-predator reflexes, and our approach to culling must take this into account.

Stalked regularly, deer become shy and eventually more nocturnal.

Our deer are relatively easy to approach at all times.

ABOVE AND RIGHT *Most census work is done from vehicles, with the deer quickly becoming entirely used to the daily traffic through the forest. Tolerance of vehicles is in itself probably the single most important objective to achieve.*

Photo by Brian Phipps

A policy of quality-based deer management can only work if constant attention is paid to habitat assessment. It is easy to forget that, as farming practices change and the forest develops through its growth stages, so too does the habitat as a source of food and cover. Just because 'X' Copse has 'always' been a wonderful corner for deer, there is no reason to suppose that it will continue to be so. Changes are slow, but over ten years may be very significant. Maintaining deer at relatively high densities therefore requires continuous assessment of the habitat, particularly where changes in agricultural cropping policy can dramatically alter the availability of vital winter keep.

Maintaining deer at relatively high densities requires continuous assessment of the habitat.

There is much talk today about the 'damage' that deer are doing. This may be damage to trees, crops, gardens or woodland biodiversity. It is generally stated that deer at 'high densities' are responsible for the damage, and figures are sometimes published as guidelines. What is very clear to me, and was once confirmed following a visit to one of my management areas by an expert in her field, is that the term 'high density' is relative. Damage in some habitats can be severe even when deer numbers are what I would think of as low (something like ten deer per sq km of woodland). However there appeared to be no visible impact on the biodiversity in the visited area, despite roe deer being present at what some plant experts would consider to be very high deer densities (something between 50 and 75 deer per sq km of woodland). Indeed, the secondary growth within the forest structure, the natural regeneration and the wide diversity of ground flora led to an initial opinion that the deer were in fact at low density. Clearly some damage is species specific, and fallow deer are the most likely species to be named as the culprits. However one inescapable message is that 'managed' roe deer even at 'high density' can be entirely consistent with low incidence of 'damage'. Furthermore, the simple equation that is assumed by some to exist between deer density and deer quality must also be amended, to take account of an infinite variety of habitat types, which are more or less susceptible to the impact of deer. This introduces a second inescapable message that, under managed conditions, 'high' roe deer numbers may be entirely compatible with 'high' quality.

So, here in the south of England, we appear to be lucky enough to manage areas which, despite relatively high roe density, must be of good enough overall habitat quality to mean that deer quality is high, and damage either to forestry, agricultural crops or natural biodiversity has not really ever been an issue. Furthermore, it must be remembered that agricultural damage by roe and fallow deer is usually negligible when compared to that of pigeons, pheasants and rabbits. Even 5 per cent of agricultural yield loss (and research shows that 1–2 per cent is more likely) pales into insignificance when compared with annual price and yield differentials of up to 40 per cent based on deficits of weather, husbandry and crop prices. Deer damage in woodland is often negligible when compared to losses through problems of ground preparation, of plantation maintenance, of damage by other mammals such as sheep, squirrels, voles, rabbits and hares, and, of course by insects. Nevertheless, we now assume that no one today expects to plant unprotected trees. Were it, however, the policy to do so, then I suspect that we might have to reconsider our woodland planning. In other words I have been managing deer where they are considered to be something of a priority, and this is bound to have had some impact on the potential of our policy to succeed.

Agricultural damage by roe and fallow deer is usually negligible compared to that by pigeons, pheasants and rabbits, and loss to factors such as adverse weather.
Photo by Brian Phipps

A good deer manager, like a good game-keeper, must have no consuming personal interest in shooting. Of course the deer manager must be capable of getting out and achieving sometimes quite significant culls, but must not be driven principally by a desire to do so. Any aim that a landowner might have for quality deer must therefore be balanced by employing a deer manager who is not personally interested in stalking as sport. The stalker who measures success simply by the ability to cull large numbers of deer may have a place in some operations, but is not necessarily the right employee for the hands-off approach which suits a low-stress, high-quality management technique. The adage is to leave the deer alone as much as possible.

The wonderful thing about deer is that two people, studying the same species but in different areas, can make entirely contrasting deductions about their behaviour. What I set out in this book is what I have learnt so far from almost 25 years of full-time professional involvement with deer in Hampshire, principally with roe and fallow. Thus the opinions expressed are personal, but supported by the evidence gained from other respected professionals, to whom I am very grateful. I am aware that I still have much to learn, but I am also aware that the knowledge that many professionals have gained over as many years is widely ignored by the new decision-makers who, following a brief association with stalking as sport, feel qualified to speak on behalf of the deer and of their management. The professional's way may not be the science of the classroom, but is perhaps all the stronger for its empirical foundation.

2.

CULL PLANNING

MUCH THOUGHT MUST be given to planning how many deer are to be culled. With the benefit of many years experience of roe and fallow over the same area, setting that figure poses few problems. The historical cull figures can be analysed very easily to see what affect they have had on numbers and sex ratios, and then adjustments made following the current season's census. Provided that the spring census is carried out under established standards each year, then trends in the population are soon apparent. In the mixed open woodland of north Hampshire, deer are readily observed on the woodland edge, and so direct observation is a satisfactory method of census.

My approach with roe is to identify and count as many territorial bucks as possible during a given period in March, and to mark them on an Estate Map. Where possible, a description of the antlers is made so as to enable future identification, and photographs are taken if the opportunity arises. It is not practicable to mark more than that, as any young deer will soon be dispersed, and mature females are broadly indistinguishable from each other. It is accepted that this cannot be an exhaustive count, and in practice up to one-third of the mature buck cull is made up of animals which have not previously been recorded in the spring count. This suggests that in open woodland it may be possible to count about 70 per cent of the population by direct counting methods and, taking the input figures from my own spring census, this extrapolates to a density figure equivalent to 54 roe per sq km of woodland, assuming our established sex ratio of almost 1:1. In explanation therefore, take the recorded number of mature roe bucks at the spring count (say 45), multiply it by 1.5 according to cull returns which suggest in this case that only two-thirds of the mature bucks are observed. Thus, in a population with at least 67 territorial bucks with a balanced age structure, an estimated 67 young bucks must exist alongside. In a population of even sex structure, as observed in our spring count, 134 bucks suggests 134 does and thus a total minimum population of something like 268 roe. If the total area of woodland on the Estate is 500 hectares (1,236 acres), density can then be expressed as 54 roe per sq km of woodland.

ABOVE AND LEFT *My approach with roe is to identify and count as many territorial bucks as possible during a given period in March, and to mark them on an Estate Map. Photographs are taken if the opportunity arises.*
Photos above by Chris Howard

Photo by Brian Phipps

In open woodland it may be possible to count about 70 per cent of the population by direct counting.
Photo by Chris Howard

The methodology of counting remains the same each year. The count is undertaken at the same time of year, by motor vehicle, and using the same established route. Using two clickers, sightings are recorded of the total number of males and females seen. This significant sample method gives you your resident sex ratio, and is easily documented for future comparison. Similarly, average and maximum outing counts are recorded and again are readily referred to for future comparison. Thus by counting and recording the total number of territorial males, and by comparing the average and maximum outing counts, you can readily establish whether numbers have gone up or down since the last year. Of course there will be anomalies, usually associated with the specific weather conditions during the census period, and with that in mind I have always sought to choose a week of 'average' conditions for the time of year. By counting your territorial bucks in this way (and remember, territories remain pretty well unchanged each year), you can be absolutely sure not to over-exploit the most valuable and vulnerable class of animals within your population.

The fallow census is based on observed numbers and sex ratios only, and is not accorded the same perception of accuracy as is believed to be possible with the roe. During March, however, I would expect each group seen to be discrete so it should be possible to avoid double-counting.

Remember that territories remain pretty well unchanged each year.
Photo by Chris Howard

It is nevertheless important not to get carried away with trying too hard to establish exact density figures. Using any method ever devised, it is impossible to make an exhaustive count of wild deer. However, while estimations are an important part of cull planning, and may be more or less accurate, of crucial importance is assessing *trends*: have deer numbers appeared to rise, fall, or stay about the same, and are the sex ratios satisfactorily maintained to as close to 1:1 as possible? What is your neighbour doing, and does there appear to be any net import or export from or to neighbouring land? There are any number of methods of assessing deer populations other than by direct counting but, with respect to 'visible' deer, these are largely unwarranted. I have used thermal imaging, which is frequently held to be a panacea in deer counting, and found it to be a useful supplementary method. I was nevertheless finding more activity and higher counts using normal dawn and dusk observation, and on one occasion we saw no deer at all during a night patrol. I have never tried dung counting, but again if the enormous resources required were available, it could produce useful supplementary information particularly in respect of 'unseen' deer in dense woodland. But when all is said and done, unstressed deer are readily observed and a reasonably accurate assessment of the population becomes possible.

In summary, using a system of average and maximum outing counts and recording the observed sex ratios during the spring census will give a

ABOVE *Are the sex ratios satisfactorily maintained to as close to 1:1 as possible?* Photo by Chris Howard

Unstressed deer are readily observed and a reasonably accurate assessment of the population becomes possible.

reasonably accurate assessment of the population, and a very reliable indicator of annual trends.

It is also possible to estimate your population by using others' experience of average density figures, which have been derived from cull returns provided by professional deer managers. Successful deer managers in lowland conditions take annual culls of between 13 and 22 roe per annum per sq km of woodland, and this has proved to be adequate to keep the population stable. On the basis that they are probably consistently correct, then density levels of between 43 and 75 roe per sq km of woodland (spring population) must exist in average habitat conditions in the south. There is, of course, considerable variation according to habitat, and it is therefore important to have a good understanding of what constitutes 'good' habitat for deer, and what 'poor'. A mixed age-class, mixed species patchwork of small plantations amid productive agricultural land, where winter sown crops are abundant and game crops planted for pheasants, may create habitat holding up to 75 roe per sq km. An old beech wood on the top of a windy ridge, with little ground cover or understorey, may hold fewer than ten per sq km.

While an estimate of total numbers is an important foundation to cull planning, the balance of sex ratios is absolutely crucial. Although it is generally taught that even sex ratios are an unattainable ideal, observation suggests that a rigorous culling policy of shooting up to twice as many does as bucks over many years has achieved this position, at least with the roe. Just consider the breeding potential of a population of even sex ratios, as against one with a 70:30 split in favour of females. In a population of 100 roe, the birth rate could rise from perhaps 75 to 100 in average habitat, or even more in the best habitats. It will probably not be possible to achieve the same balanced sex ratio with respect to fallow, as the nomadic nature of the bucks tends to result in losses outside your control.

The balance of sex ratios is absolutely crucial.

Does, even to the trained and experienced eye, look pretty similar and even the very old are sometimes difficult to identify.
Photo above by Chris Howard

In addition to total numbers and sex ratio, the spring census should provide valuable information on age-banding within the population. However, it is important not to get carried away by the science. Does, even to the trained and experienced eye, look pretty similar and even the very old are sometimes difficult to identify. Nevertheless it is possible to allocate does into 'mature' and 'yearlings' (last year's kids becoming 'yearlings' in March). You should also have in mind that, within a well-structured population, it is desirable that about 30 per cent of your adult roe (20 per cent with fallow) are yearlings, in order to provide replacements to the pyramidal population profile which exists for all prey species. Thus it is important that there should remain sufficient doe kids after the winter cull to provide that 30 or 20 per cent. Having mentioned the word 'pyramidal', I cannot avoid reference to the oft-maligned 'Hoffman Pyramid'. Although it has, in the past, been advocated as a detailed cull-planning tool, it is better to understand it simply for what it teaches us about the *structure* of cull planning. The foundation of the concept is that, in an unmanaged population of deer, a cohort (or birth year, i.e. all the kids born in a particular breeding season) will suffer annual natural losses over their life-span that will eventually lead to a large number of young deer becoming a small number of old deer. Thus, to reach a situation where deer are able to reach their full age potential, it will be necessary to support a population which contains progressively larger numbers of young deer. The pyramid assumes an even birth ratio between males and females, and the summit of the pyramid represents the average maximum age of the species.

Last year's kids becoming 'yearlings' in March.

In a naturally predated population, the overall size of the pyramid should remain stable. Here in the UK, where we have removed all the natural predators, the pyramid will tend to enlarge. Figure 2.1a shows the diagrammatic pyramid, to which the detail of a typical population is added in 2.1b. In 2.1c the summer births are added, and the unshaded boxes in 2.1d represent the annual potential surplus to the population

The Hoffman Pyramids

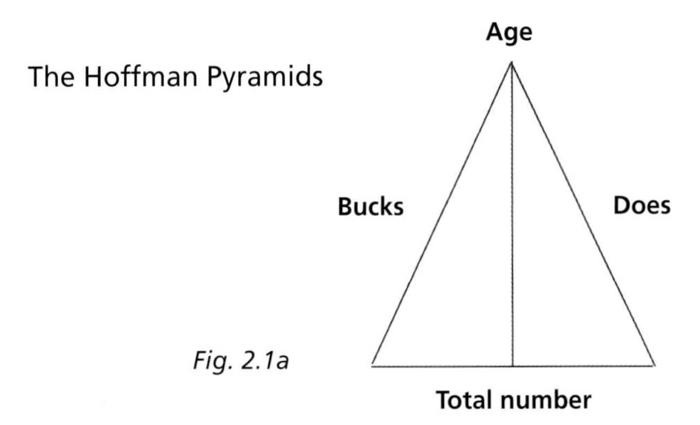

Fig. 2.1a

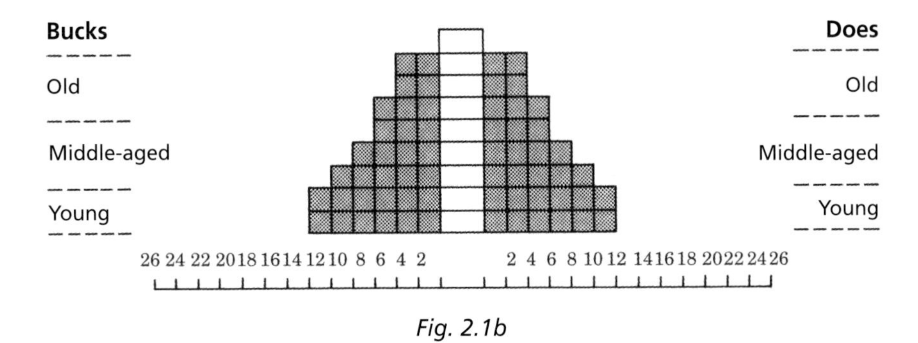

Fig. 2.1b

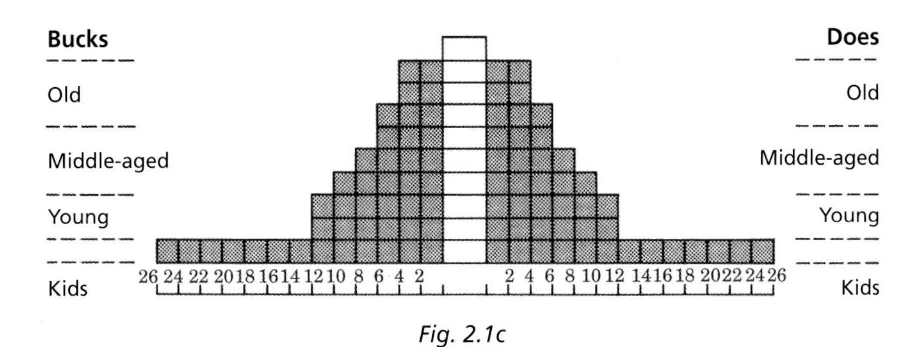

Fig. 2.1c

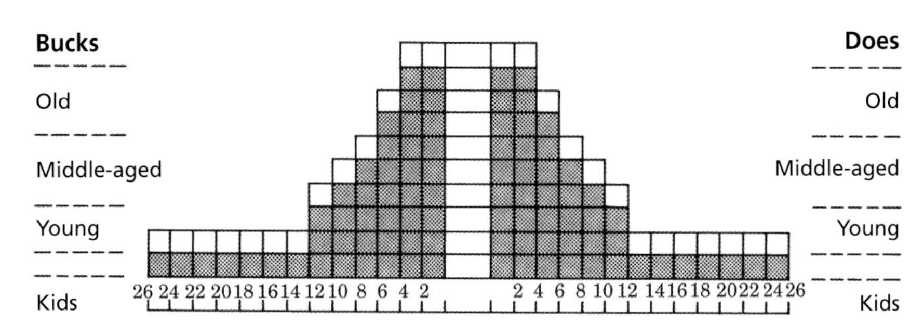

Fig. 2.1d

before the effects of mortality. Thus the purpose of deer management is to emulate natural predation, and simply remove that representative slice from the population.

This introduces the next variable, which is to have some knowledge of your breeding rates, which can be calculated or observed. By collecting the uterus from each doe culled after Christmas and counting the foetuses present, it is possible to estimate the birth rate from the population as a whole. In the case of roe, do not make the mistake of attempting to count *corpora lutae* in November/early December before the commencement of the development of the foetuses. Because of the effect of embryonic diapause (delayed implantation) the presence of *corpora lutae* does not indicate a pregnancy, merely the fact that the doe has ovulated. If, however, you complete the bulk of your cull before Christmas, this method of assessing breeding potential will produce too small a sample to be useful, and you will have to rely on simple observation post-fawning.

Remember that there are two classes of doe present during the spring census: mature does which will almost certainly be pregnant, and yearlings which will definitely not. Some scientific research has suggested that, in good habitat, roe kids will be successfully impregnated during their first rut when they are just 8–12 weeks old. I have to say that, in all of my experience over many years, and in the experience of several other full-time professional deer managers, we have never come across a single pregnant kid. The most likely explanation for the research is that the researcher mis-aged 18-month-old does as 6-month-olds, being unaware of the differences in tooth eruption and development between roe and red deer. In any event it is important to define how you will express your birth rate, whether as a percentage of the 'breeding' females, or as a percentage of 'all' females including those not yet of breeding age, and it is considered normal practice to use the latter, not least because that makes the overall calculation easier. In my area of northern Hampshire, most roe does have twins, and I estimate the overall birth rate to be as much as 1.8 kids (180 per cent) per *breeding* doe, which equates to a figure of about 1.5 kids (150 per cent) per *total number of* does in the spring population. In more challenging conditions of the uplands, birth rates will be nearer 1.4 kids (140 per cent) per breeding doe, which equates to a figure of one kid per total number of does in the spring population – i.e. a birth rate of 100 per cent of the total female population. Of course, this figure of one kid per doe makes the maths of estimating the birth rate very easy. For fallow, a birth rate of

In the south, most roe does have twins.

75 per cent of the total number of does within the spring census might be appropriate.

The remaining variables to the equation are mortality, and emigration/immigration. Mortality rates are extremely difficult to assess. Sample counts just post-fawning compared to sample counts on the autumn stubbles may give you an estimation of neo-natal mortality, but adult mortality is very much more difficult to work out. Deer that die in the woods are rarely found, the carcasses usually being devoured by badgers within days. Even motor accidents cannot be recorded in full, as more and more motorists now appear to know the value of fresh venison (legal or not)! Perhaps the easiest way to account for mortality within a cull plan is to make assumptions of say 10 per cent in adults and 25 per cent amongst kids (i.e. neo-natal mortality), only changing these figures if particular circumstances dictate that amendment is necessary.

Immigration and emigration can be the Achilles heel of any cull plan. Even if numbers appear to have remained stable from year to year, it might be despite your own under-culling which has resulted in emigration, or despite your own over-culling resulting in immigration. However, we have to assume that our deer management is broadly effective, particularly when seen over many years, and the most pragmatic approach is to assume that the effects of immigration and emigration are broadly neutral, and thus ignore it unless your particular circumstances dictate otherwise.

Thus the key population factors affecting cull planning are:

1. Overall density.
2. Sex ratio.
3. Age-class structure.
4. Reproductive rate.
5. Mortality.
6. Emigration/Immigration.

Having considered all the above factors, a cull plan must be developed. One way is to simply apply the '30 per cent rule'. This is based on an understanding that a cull of 30 per cent of the spring population of roe represents the net annual recruitment to the population. This figure will, of course, vary according to regional and local conditions, but nevertheless seems to work as a useful practical guide. Established British Deer Society guidelines, which have withstood the test of time, suggest that the total cull should amount to about 30 per cent of the estimated spring roe population, of which 60 per cent should be female and just 40 per cent should be male. Mature animals should make up 40 per cent of the cull, and yearling/kids 60 per cent. Although these guidelines represent achievable and practical best practice goals, I have refined them by further reducing the number of bucks included in the cull, simply to make up for a general imbalance in the country as a whole caused by regionally disproportionate culls of males. Thus, simply leaving a few extra young males, despite the fact that they may well be exported to our neighbours as a result, may be a conservative and future-proof policy. Remember that they can only be shot once, and each one left has at least a chance of surviving a few more years. My own cull figure for roe averages over 20 years at 67 per cent female and just 33 per cent male. For fallow, because of the marauding nature of the bucks, I limit this further to a cull of about 80 per cent female and just 20 per cent male.

A more detailed cull-planning model is printed as Figure 2.2a for roe and 2.2b for the lowland herding species. While this model can be used for practical purposes, it is particularly useful to support, explain and understand the concept of a structured cull, and is therefore beneficial when attempting to plan a manipulation of the population. Those with a fear of maths will no doubt fight shy of it, but it is designed as a flow chart and is reasonably straightforward to follow. The multiplying factors are inserted

Photo opposite by Chris Howard

The total cull should amount to about 30 per cent of the estimated spring roe population, of which 60 per cent should be female.

to make it even easier, but consideration must first be given as to whether these multiplying factors are actually appropriate to your particular piece of land and, if not, they should be amended accordingly.

For example, in respect of roe, first complete Rows A and B that represent the spring census, which may be observed or estimated. Row C divides the adults into mature and yearling and adds the number of kids estimated to be born during the summer. Then jump to Row M and decide how many deer you would like to see this time next year, and in Row L further refine your requirement to address any imbalance in sex ratio. Working now up the page, divide your females and males according to a structured age grouping (remember that with roe about 30 per cent of your adults should be yearlings). Row J simply recognises that *next* year's adults derive from *this* year's adults, *plus this* year's yearlings, and mirrors the required 30 per cent yearling age structure. Furthermore *next* year's

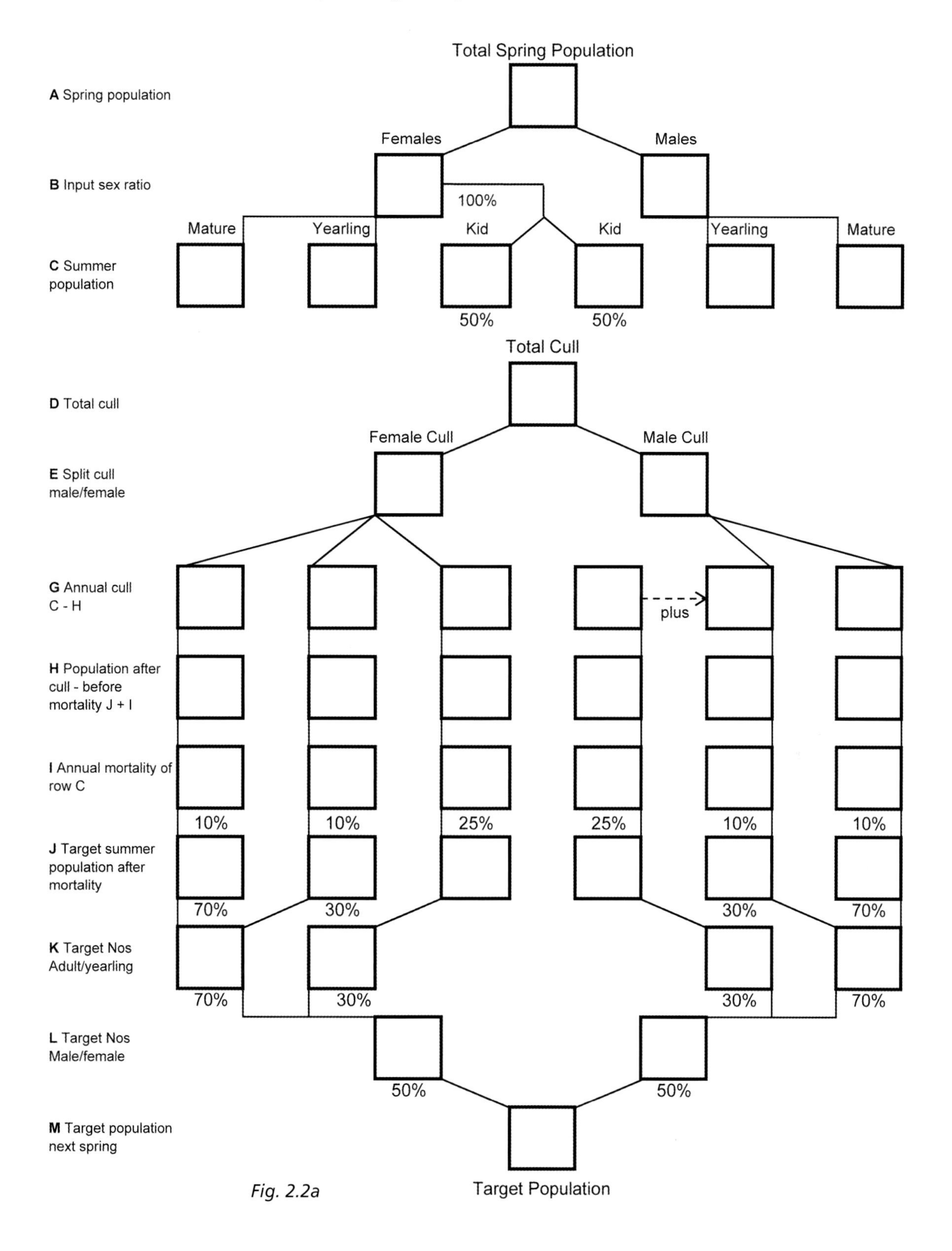
A cull-planning manipulation model – roe deer
Total Spring Population
A Spring population
Females
Males
B Input sex ratio
100%
Mature
Yearling
Kid
Kid
Yearling
Mature
C Summer population
50%
50%
Total Cull
D Total cull
Female Cull
Male Cull
E Split cull male/female
G Annual cull C - H
plus
H Population after cull - before mortality J + I
I Annual mortality of row C
10%
10%
25%
25%
10%
10%
J Target summer population after mortality
70%
30%
30%
70%
K Target Nos Adult/yearling
70%
30%
30%
70%
L Target Nos Male/female
50%
50%
M Target population next spring
Target Population

Fig. 2.2a

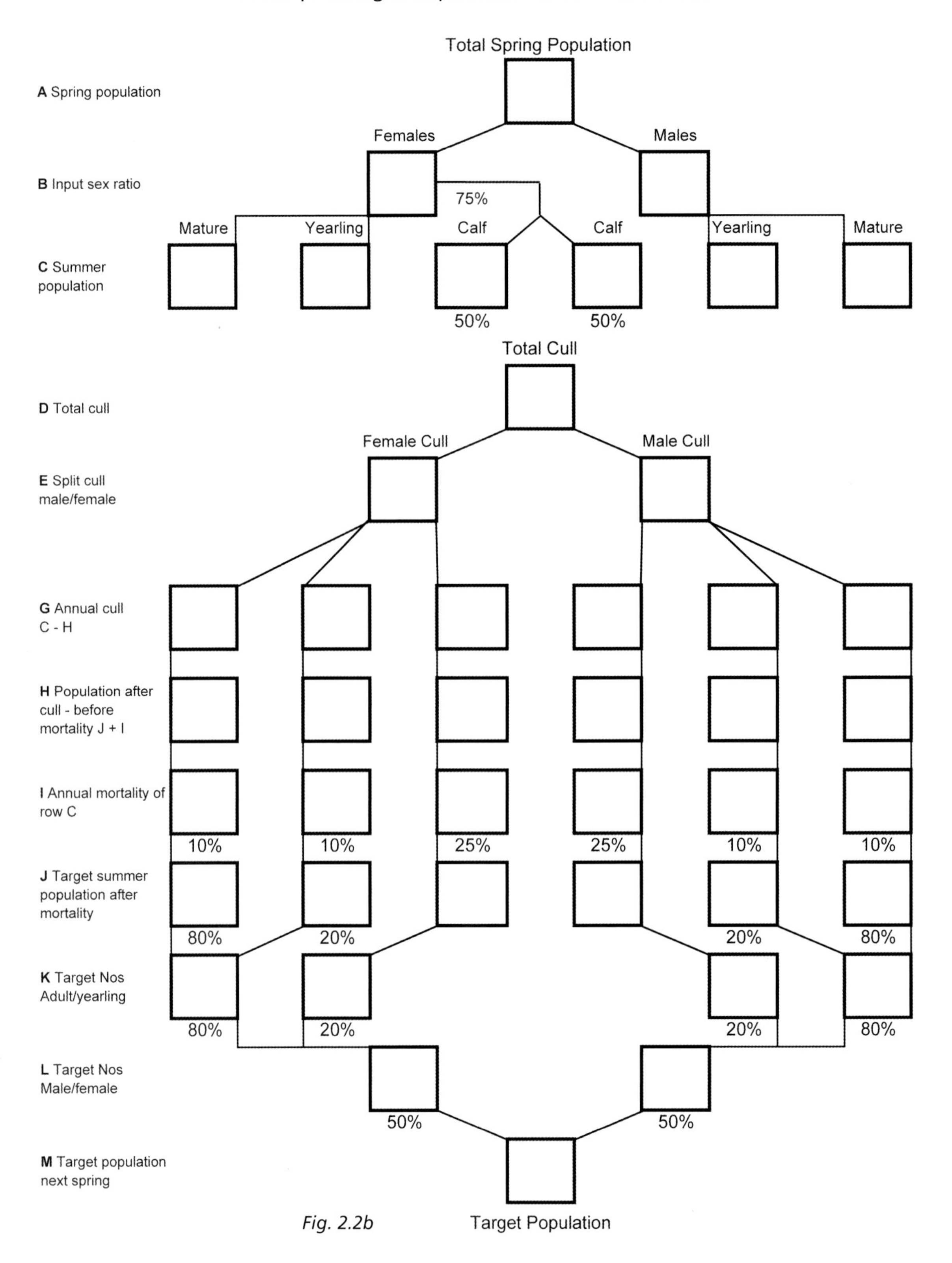
A cull-planning manipulation model – fallow deer
Total Spring Population
A Spring population
Females
Males
B Input sex ratio
75%
Mature
Yearling
Calf
Calf
Yearling
Mature
C Summer population
50%
50%
Total Cull
D Total cull
Female Cull
Male Cull
E Split cull male/female
G Annual cull C - H
H Population after cull - before mortality J + I
I Annual mortality of row C
10%
10%
25%
25%
10%
10%
J Target summer population after mortality
80%
20%
20%
80%
K Target Nos Adult/yearling
80%
20%
20%
80%
L Target Nos Male/female
50%
50%
M Target population next spring
Target Population

Fig. 2.2b

yearlings derive from what remains of *this* year's kids. Row I makes some allowance for mortality, which can be estimated, calculated or observed, but must always relate to the spring population (Row C). Row H is a sub-total which adds back the mortalities in order that the cull figure can be calculated. Row G represents the planned cull in detail and is simply the difference between Row C, the summer population, and Row H (what must remain before the influence of mortality). Rows E and D total up the culls to give a grand total figure. Note that in row G, even though male roe kids may now be legally culled during the winter following the Regulatory Reform Order (Deer) 2007, the option remains to carry forward this cull as a cull of yearling bucks in the following spring.

Do not allow the maths to get the better of you to the extent that the message is lost. Look at each cull figure and ask whether it is practical and achievable. Experimenting with this model will demonstrate that only very minor manipulations can be made in a single year. Detailed long-term deer management plans are completely impractical, and such plans should therefore only ever reflect broad aspirations, which will require more detailed analysis on an annual basis depending upon what the

Only very minor manipulations can be made in a single year.

spring census suggests has been achieved. If the overall cull figure in Row D represents a cull of in excess of 40 per cent of the spring population, then the plan should be abandoned as undesirable, in that it will have a destabilising influence on the deer. If the plan suggests a disproportionately large doe cull, it may well be unachievable and should similarly be abandoned in favour of a more modest plan. Remember that this is only a 'plan', and plans rarely survive the first shot. Nevertheless, they are an important part of structured deer management, and often serve as a check on what your intuition as a stockman suggests. When all is said and done, however, the 30 per cent rule is often enough.

3.

ACHIEVING THE DOE CULL

PLANNING A CULL IS ONE thing, completing it successfully is quite another. Do not forget that for all the planning in the world, it is equally important to be able to assess what is actually happening on the ground, and to be able to be reactive to any apparent deficiency in your plan as it arises. In terms of qualitative deer management, I have always taken the view that it is counterproductive to over-stalk the deer, and one of the worst possible scenarios is where a permanent member of the Estate staff is instructed to cull deer on his rounds. To the Estate Manager it is seen as a good use of time and resources, reducing the costs of additional labour and machinery. But they forget that deer are a prey species and have strong anti-predator instincts. In that instance, therefore, the deer become habitually stressed, associating every movement in the forest with potential danger, and furthermore soon becoming shy of all vehicles. They don't go away, but they do get a lot more difficult to manage. If any member of staff is to take a part in the cull outside his normal work, then it should be for specifically allocated periods only, and on foot. A full-time deer manager will always be intensely aware of his deer and of their reaction to him. Areas will therefore be stalked intensely, and then left alone, allowing the deer unstressed enjoyment of their territories for most of the year.

There are three principal ways of achieving the cull: first, through stalking and second, through the use of high seats (about which a very great deal has already been written over many years). The third way is

It is counterproductive to over-stalk the deer.

BELOW *Deer are a prey species and can soon become habitually stressed, associating every movement in the forest with potential danger, and soon becoming shy of all vehicles.*

through gently moving the deer, which is essentially a combination of the previous two. If the winter cull is to be of any immediate benefit to the herd as a whole, then culling should start as early in November as possible, thus removing the surplus before their impact on the dwindling food supply will be apparent. Although it is clearly impossible to achieve the entire cull before the New Year, I always tried to complete 60–70 per cent of it before Christmas. The deer are relatively active in November, and success by stalking and sitting comes readily. The key is to keep an eye on the stubbles in mid-October, because the deer using them then will continue to use them up until about mid-November. This will ensure the best use of time right from 1st November when the season opens. By December, however, the onset of winter inappetence begins to affect the behaviour of the deer. They become much less active, and therefore much more difficult to stalk. I therefore set aside specific weeks to cull individual beats, stalking intensively for that week, and then leaving it alone for at least a month. In this way the majority of the cull is achieved early, with some fine-tuning possible later in the winter. However, strong forces work against you; the overarching interest of the pheasant shoot will frequently frustrate deer stalking. Keepers want peace and quiet at that time of year and though many stalk themselves they are often loathe to allow access to other stalkers until the season ends in February. There is, of course, a middle ground. The stalker must be sensitive to the Shoot, avoid stalking on the days leading up to a shooting day, and tolerate any time or area restrictions put on him, particularly around the release pens. The fact is that most of the concerns are in the mind and a stalker up a seat or taking a single shot whilst stalking has no effect at all on the success of a day's shooting.

Moving deer

An alternative, or supplementary, method of achieving the doe cull, particularly in forest situations, is to move the deer gently to seated Rifles. This is a minimal-stress method of managing deer which, if done properly, is a humane and efficient way of completing the doe cull very quickly. The concept is that instead of one person sitting out for say ten evening or morning stalks, you place ten people out for a single daytime outing of about three hours. You then add a few walkers, typically just one per 40 hectares (100 acres) of woodland, to stalk gently through the forest, lifting the deer from their daytime couches and causing them to move along their regular racks. As the deer move, alert but not alarmed, they might pass one of the seated Rifles and, if they pause long enough, might give the opportunity for a shot. The number of walkers will depend upon how thick the cover is, and how quiet the deer are. If the deer are flighty or the forest very open, then one per 80 hectares (200 acres) might be sufficient. If the deer are rather tame or the forest rather dense, then even one walker per 40 hectares (100 acres) might not be sufficient to get them on the move. Every care must, of course, be taken to avoid panicking the deer. An exercise of this type can be undertaken at dawn or during the day, but not in the evening when carcass collection, which is always carried out after a movement, will become a problem.

The deer are not driven – indeed, if they break into a run, then the situation is lost and you might as well go home. If, however, in their ambling, the deer happen to stop in front of one of the seated Rifles, then he should have plenty of time to identify, select and shoot if appropriate. Indeed, properly executed, the success rate from this method may result in roe or fallow culls of between two and four times the number of Rifles involved. This represents an average success rate even greater than that of a full-time professional stalker, let alone the recreational stalker who typically achieves just one cull per three outings.

Movements will not work if done regularly, and should be restricted to just once or twice per winter. Success is readily achieved in February, and game shooting commitments on most Estates restricts its use at other times of year. The key to success lies in the following factors.

Rifles

The number of Rifles will depend upon the lie of the land, the density of deer and the requirements of the cull. Too many will fill the forest with scent, and unsettle the deer, but a figure of one Rifle per 20 hectares (50 acres) of woodland seems to prove successful. Placement of the Rifles is of equal importance. The absolute minimum of human disturbance, time and vehicular movement should be used in delivering them to their seats, and seats should all be sited adjacent to forest tracks so that there is no requirement for the Rifles to spread scent around the forest through walking to their seats. The forest should then be left undisturbed for half an hour. During that time some movement will almost certainly take place and the cull may begin. In some instances, particularly with fallow, the amount of spontaneous movement has been sufficient that the walkers never needed to enter the wood and the cull has proceeded entirely without intervention.

The Rifles must all be of greater-than-average skill and self-discipline. There are many who are keen to offer to help, but not as many with the advanced skills necessary to identify and shoot, quickly and with confidence. Rifles must have access to communications, and be briefed as to the timing of the operation, the direction of the movement, the location of all Rifles, the age-class division of the required cull and of course the location of any public roads or tracks. Collection and delivery procedures must also be given careful thought.

Walkers

The key to success lies in the walkers' thorough knowledge of the woodland, of the deer, and of how the deer use the land. Game shooting 'beaters' are entirely inappropriate unless thoroughly re-briefed and genuinely enthusiastic to take part. The objective here is a quiet approach or stalk which 'alerts' rather than 'alarms', and which works 'around' the deer rather than 'at' them. Most of the work is done by the wind carrying the scent of the walkers to the deer long before they can see or hear them. For that reason movements should always be undertaken downwind. It is generally the case that stalkers make the best walkers, and in many cases the most experienced within the team of Rifles would actually be better employed for the day as a walker. The walkers should also know exactly where on their route each Rifle is sited, and can then avoid approaching too near.

High seats

For normal stalking purposes, high seats are usually sited on field edges, overlooking the regular deer feeding areas. The concept is that the seat can be approached by the stalker well before the deer come out to feed, and without the risk of disturbing them from their woodland couches. When moving deer, however, such seats may be inappropriate as the deer are inevitably less relaxed when approaching or crossing open areas. Instead, seats should be re-sited according to the deer's known movement pattern to areas of cover dense enough for the deer to feel relaxed, but with enough visibility to select and shoot with confidence. This entails a great

The objective is a quiet approach or stalk which 'alerts' rather than 'alarms', and which works 'around' the deer rather than 'at' them.

deal of extra reconnaissance and physical labour, but success can only be guaranteed if every single seat is assessed in this way, and indeed re-assessed each year as the deer's use of habitat adjusts. In the case of fallow deer, there will also be key fields to where the deer will relocate once disturbed in the woods. Seats overlooking these fields can be very useful, but a tight group of milling fallow can produce impossible shooting conditions for all except the most experienced Rifles.

Safety

There is no reason why this method of culling should add any burden to the Estate's normal Risk Assessment, although it is prudent to consider it as a separate operation for Health and Safety purposes. All Rifles are experienced individuals, shooting only from high seats at stationary targets with solid earth backgrounds. With specialist briefings, communications between all parties, formal emergency procedures, and sensible additional precautions such as fluorescent jackets for the walkers, movements can be carried out safely and efficiently. At best, therefore, moving deer is an invaluable management tool which has particularly beneficial welfare implications, as it is possible to start and complete a cull sometimes in a single day, thus leaving the deer completely unstressed for almost the entire season.

Using this system, four of us used to effect a selective cull of 60 fallow does in three to four days, with three sitting and one walking, this on an area where a single stalker on his own would struggle to shoot one or two in a day. On another area the same number of fallow are shot by a slightly larger team in a single operation of four hours, leaving the deer completely unstressed and back to their normal relaxed behaviour by evening. If roe are included in this type of procedure then it is possible to start and complete a cull in a day. It is an extremely efficient form of culling and one which, if better understood, could have made the call for extended seasons completely unnecessary (see below).

A European perspective

One of the most interesting jobs I have ever been given was to advise a German owner of some 8,100 hectares (20,000 acres) of forest. Unusually, the forest had a population of wild boar which lived alongside an important population of fallow deer. As is usual in Europe, the cull was taken in a series of driven days. However, they were experiencing significant problems owing to the pressure of 'driving' which meant that whilst the boar behaved in a predictable way, the fallow would run past the Guns in full flight, causing difficult shooting conditions and associated welfare problems. The trouble is, that when a boar runs it runs 'flat', and is actually not half as difficult to shoot as one might imagine. A deer, however, always bounds and, if running, provides an almost impossible rising and falling target. Strange though it must seem for an Englishman to be advising Germans on matters of 'hunting', that was what I was asked to do based on my experience of moving fallow in the UK. I visited the area and examined the habitat; it was not so very different from UK woodlands. My basic recommendations were as follows:

1. Deliver the Rifles and wait half an hour before starting the drive.
2. The drive should be conducted downwind.
3. The beaters should make no noise and, if possible, walk in two waves.
4. Dogs should only be used in the last half hour of the drive.
5. Move seats off rides (a process already under way with the new head keeper).
6. Re-brief the Rifles to expect only walking deer, which they would have time to select.

It is easy enough to make recommendations! However, they accepted every one of my proposals and undertook to trial them. What is more, I was invited on the trial day.

The fact that it was a success was an enormous relief. Thirty shots were recorded before the drive even began. Deer had crossed rides and slowed down in the semi-cover where the seats were sited. No fallow were wounded and, much to everyone's surprise, the boar moved without the need for dogs and a good number were included in the day. As a mixed day it was a huge success. I think that over the weekend a total of 70 fallow and 50 boar were included in the cull.

Does and their kids

There is much argument about how one should approach the problem of does with young in November. As I have said previously, if any real benefit to the remaining deer is to be achieved in terms of increased winter food supply, it must be imperative to commence the doe cull early. The Regulatory Reform Order (Deer) 2007 has made some real sense of the previous legislation regarding the shooting of buck kids. It is now thankfully legal to shoot a buck kid if you can demonstrate that 'it has been, or is about to be, deprived of its dam'. Nevertheless it is all too easy to see a kid as 'sexless', and forget that although it is only a kid it is still a male; a buck which, in the past, you would have culled as a yearling next spring. With this in mind it is important to record its sex and include it in your total allocation of bucks. My fear is that some stalkers will be loath to forego their usual allowance of yearling bucks and continue shooting a number despite having culled them 'already' as kids. This can only lead to further pressure on a population of bucks already under pressure in many areas.

This brings me back to the other much more significant effect of the Regulatory Reform Order – the extension of the doe season up to the end of March. This, in my view, will have damaging and far-reaching effects on the wonderful roe which we have in England. March was a vital month for deer welfare, giving them a break from stalking and giving us stalkers an important opportunity to take stock and carry out the annual census. It gave the does the peace and quiet they needed pre-fawning and allowed the whole forest to settle after a busy February. Taking away this quiet month is the biggest own-goal the deer industry has made. Certainly more deer will be shot, and evidence from game dealers is clear that they are now taking in more deer than previously (I suppose as the legislation intended), but this extra month is simply not needed for roe. Roe are relatively easy to stalk and to manage; there is absolutely no requirement for an extra month to control does. And March is such an easy month in which to shoot deer – too easy as they are just beginning to come out of winter inappetance. Indeed, it is clear that in some areas of Hampshire roe have now become scarce where before they were numerous. It was pretty clear to anyone with an interest in roe that this is what would happen, and I have no doubt that owing to the territorial nature of roe it could be possible to almost eliminate your population.

March was a vital month for deer welfare; it gave the does the peace and quiet they needed pre-fawning. Photo by David Mullen

I had this conversation with a Scottish stalker, and he put up a strong defence of the change. In his area he would never shoot a buck until May and, since spring comes so much later north of the border, he was able to utilise April for his census. Furthermore fawning happens later there than in the south, so while he agreed with me in principle, the conditions in his area meant that he was still able to offer his deer that all-important month off. Under these circumstances I have to agree with him, but here in the south I would advise anyone with a real interest in roe to stick to the old timetable. I remember when the changes were tabled, and I remember agreeing that, under certain circumstances, an extension of two weeks into March for fallow does would be useful. However, the legislation was driven by those with more negative views of deer, and who were unwilling to address their own deficiencies in deer control.

4.

ACHIEVING THE BUCK CULL

IT HAS TAKEN ME MANY years to truly recognise just how important it is to be able to age live deer, and how this has contributed to the successes we have recorded. Shortly before writing this, I looked at an area where there were concerns over diminishing quality. What I found were some really old bucks – so old that their antlers were like those of yearlings, in one instance just simple spikes, but on a head with the face of a bull! These had not been given a second look on stalking outings but were taking up valuable territories.

Selection of bucks to be included in the cull often produces fierce debate. In my opinion there are, in fact, only two important criteria that must be adhered to. First, you have got to have some idea of what is on the ground in spring, and second, you have got to ensure that the cull removes a representative slice of the population, targeting the older age classes, sparing the young middle-aged (or improving) bucks, and taking the required number of yearlings. Any personal preference based on various malformations or excess of annual antler growth is probably irrelevant in terms of purposeful deer management when compared first to numbers and second to age-classes. However difficult it is to age live wild deer (and only about 70 per cent accuracy will be achieved even in the widest description of age band), it is important to ensure that you can recognise a really old buck. In the absence of recognising great age a cull may well end up being a fairly random slice from the population, and as long as it is truly random, then it is unlikely that long-term mismanagement will occur. Unfortunately, too many stalkers set aside their mistakes, and continue the cull without amendment to the cull plan. If you have stalked buck 'A', shot buck 'B' by mistake, and found that it was not as good or as old as you had hoped, then I'm afraid that either buck 'A' has to be spared for the season, or, if absolutely definitely so old that it will die anyway, then a commensurate reduction in next year's cull must be made. The really important points are to fix absolutely the number of bucks to be culled, to try to avoid shooting the younger middle-aged bucks, and to be aware of and record any mistakes which you are bound to make.

Photo by Chris Howard

LEFT AND BELOW *Any personal preference based on various malformations or excess of annual antler growth is probably irrelevant in terms of purposeful deer management.*
Photo below by Brian Phipps

It is important to ensure that you can recognise an old (photos 1–3) buck and to try to avoid shooting the younger middle-aged (photo 4) bucks. Photos 1, 2 and 4 by Chris Howard

To a great extent, success will be rely upon your ability to make as few ageing mistakes as possible, and this is something which comes only with years of experience, and even then is not infallible. I have, nevertheless, found the most useful live indicator to be the angle of the coronet. Where the coronets are flat and close together (Figure 4.1) the buck is almost certainly young. Where the coronets are sloping in the form of a 'roof', and positioned apart (Figure 4.2), then the buck is almost certainly old. This may be difficult to see at 100m in poor light, but it is amazing how few stalkers check the bases of the antlers, but get carried away by height and bright points – which, if anything, can be very misleading.

This technique can be taken further and used as a valuable retrospective indicator of age post-cull. Having observed a buck in the field, selected it for culling as 'older' and shot it, it is essential to check as far as possible whether you were right. Apart from opening the mouth and checking tooth eruption and wear (useful at the extremities of youth and age), it is often said that the man who really knows the age is the one who boils the trophy. While a young middle-aged buck is an easy job, with the meat almost falling off after 25 minutes of boiling, an old buck can be a fearsome job, with the meat and sinews adhering to the skull and doubling the time taken in cleaning.

During the cleaning process it is possible to make an extremely useful check of age based on ossification of the central nasal bone. The Germans

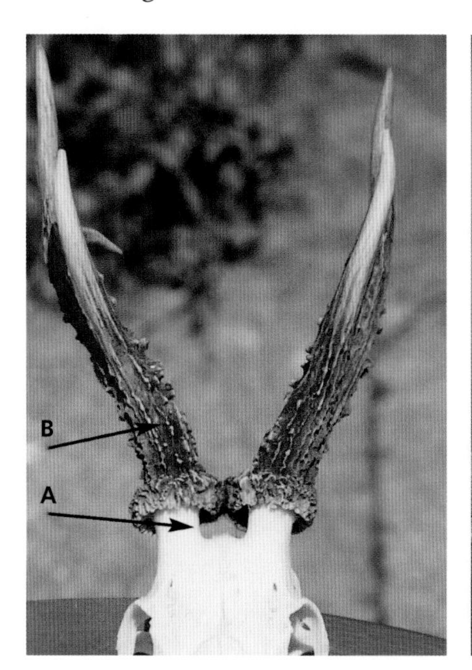

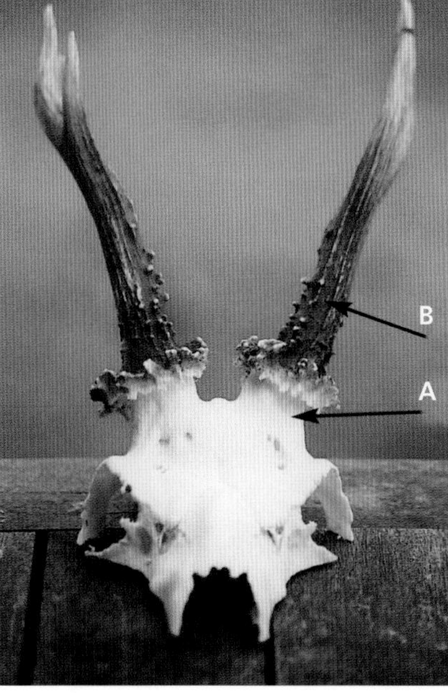

FAR LEFT *Fig. 4.1 Improving?*

LEFT *Fig. 4.2 Going back?*

have long held this method as being one of the most reliable and, perhaps unsurprisingly, it is rigidly interpreted by them. The principle is that, as the buck ages, the central nasal bone, in youth made up from flexible cartilage, becomes steadily more ossified along its length, the ossification starting from the inner part of the nasal system and progressing eventually right into the extremity of the nose. Thus a sequence of standard-cut skulls viewed from below (see Figure 4.3) shows a young, middle-aged and old deer. The Continentals take this system further, and attempt to give the deer an exact age by working out the amount of ossification as a percentage of the life-expectancy of the particular species of deer. Thus a roe with 50 per cent of its central nasal bone fully ossified might be said to be five years old. This is almost certainly an over-interpretation of a nevertheless useful ageing technique which, if used between broad parameters, has proved to be a reliable indicator of age.

Having completed the boil, there is another very useful, but perhaps less reliable, check on age which is based upon examination of the pedicle. First, the pedicle shortens with age, and second, there is a correlation between the thickness of pedicle against the thickness of antler at a point about 2cm above the coronet. In Figure 4.2, the coronets are apart and in the form of a 'roof'; the pedicle is short, and at point 'A' much thicker than the antler at point 'B'. This suggests that the pedicle might have supported much bigger antlers in previous years, and is a good indication of a buck which is 'going back', and therefore completely appropriate for having been included in the cull. In Figure 4.1, however, the pedicle is longer, and the coronets are upright and close together. The thickness of the antler at point 'B' is much greater than that of the pedicle at point 'A', and suggests a young middle-aged buck, which is either at its prime or still capable of improvement. Perhaps easy to say in retrospect, but this buck might have been better left.

However, all this theory, in which I honestly have great faith, is only that and we very rarely get an opportunity to prove our estimates. On one Estate however we were lucky enough to have a marked buck. When we first saw him he was either a yearling or a two-year-old, and he had a pronounced white spot on his right rump, giving the imaginative name of 'Spot'. We decided that, come what may, we would leave him unless we were unlucky enough to only get a view of his left side! Over the years he never really moved, occupying a central territory on the Estate and enjoying excellent food. He never grew a large head and we eventually shot him

Ageing deer from skulls

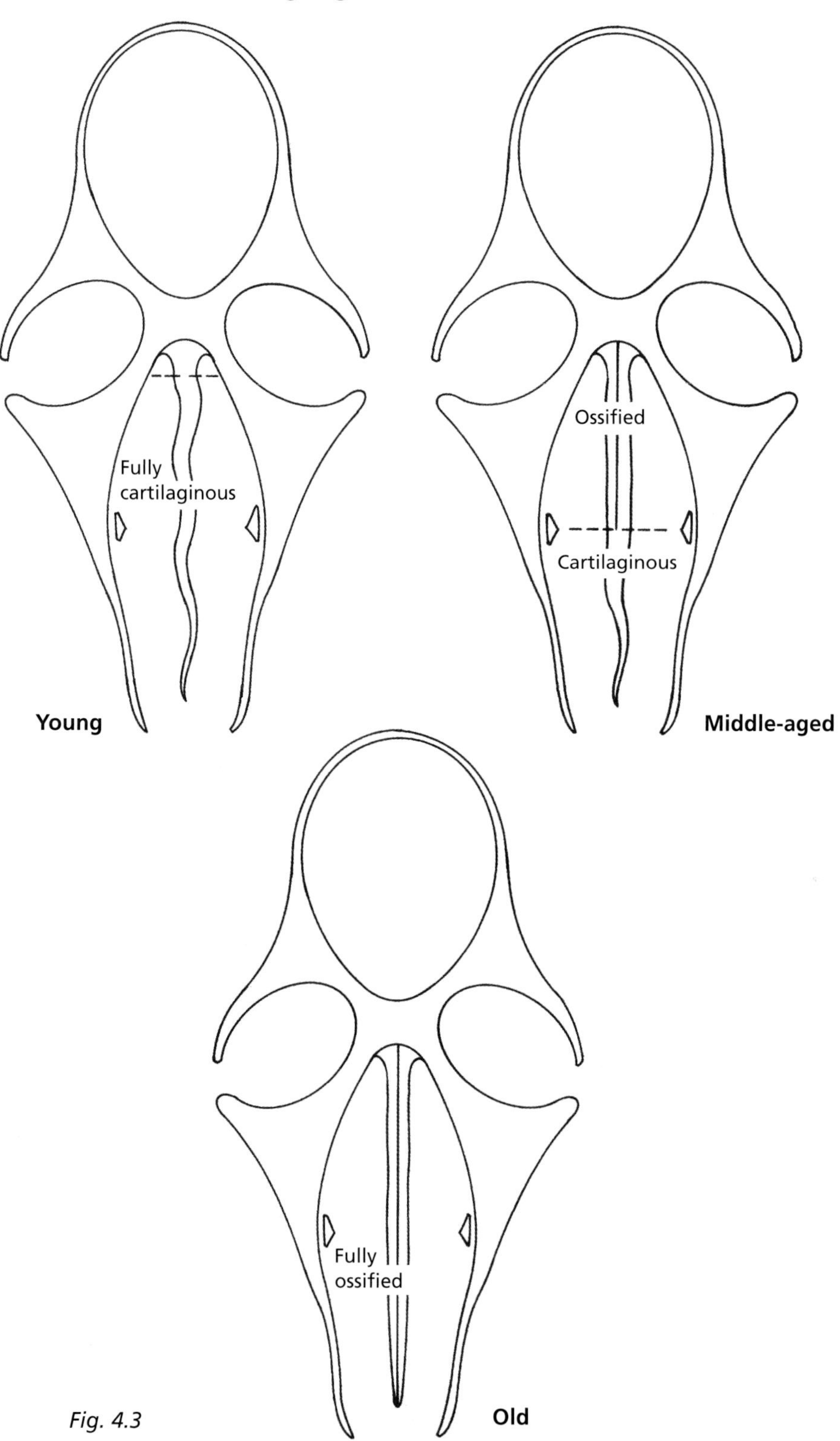
Fully
cartilaginous
Young
Ossified
Cartilaginous
Middle-aged
Fully
ossified
Old

Fig. 4.3

when he had to be seven because we were worried that he was now getting so old that we might miss the opportunity. Everything about him surprised us – no sloping coronets, no thick, short pedicles and insignificant tooth wear! (See more photos in Chapter 8.) If we had not known him to be at least seven we would have estimated him at four or five. I am not sure what this tells us about ageing, as he is the only such marked buck I have ever seen. However, what we like to think is that it proves that many of the bucks included in the annual cull are actually much older than we thought and that the herd is therefore living to its full potential.

Returning to the cull plan, at the beginning of the season I have usually identified a few of what appear to be really old bucks (say seven–ten years old), often with confusingly poor heads. This assessment thankfully often proves correct, but this represents perhaps only 10 per cent of the total cull. With at least half the cull being made up of clearly identifiable yearlings, this still leaves 40 per cent of the cull which are assessable as being say between two and six years old, i.e. middle-aged. Success therefore lies in one's ability, or just sheer luck, to shoot more at the six-year-old end than the two-year-old end. However, do not forget that even the really *obvious* ones at either end of the age range can prove confusing. Once, in April in a long, narrow wood, I was sitting in a seat at one end, whilst

He had a pronounced white spot on his right rump, giving the imaginative name of 'Spot'

400m along my colleague was sitting in another with his guest. Quite early on, a very slender (young!) buck in full summer coat (young change first, old retain their winter coat until early May) ambled past. It had a thin six-point head, and I whispered to my guest 'good yearling', and he agreed. It was walking towards my colleague, and shortly afterwards there was a shot. I was apoplectic with rage, but his guest had shot one of the oldest bucks ever seen. And really, there were no drooping coronets, no

RIGHT *I have usually identified a few of what appear to be really old bucks, often with confusingly poor heads.* Photo by Chris Howard

BELOW *At least half the cull is made up of clearly identifiable yearlings.*

rough forehead, and no thick neck, indeed nothing to suggest age. Why it was shot remains a mystery, which, under the circumstances, was not necessary to pursue.

The census work in March is essential in gauging the total number of mature bucks present on the Estate, and establishing a finite cull figure. You can never expect to count all the territorial males, and cull returns suggest that I only ever see something like 70 per cent of the total population anyway. Nevertheless, each spring I produce a map of the Estate, showing the presence of each clearly identifiable mature territorial buck, and annotate those that I would be happy to include in the cull, and those that appear too young. There is no point in including the yearling males on the map, as they are non-territorial and will soon be evicted by their mothers, and may even leave the Estate. The picture built up over the years demonstrates whether there appear to be more or fewer mature bucks in a particular season, and the cull plan can be based upon that assessment. Typically, about 45 mature territorial males are counted in one of the forests each year, and the cull plan allows for a total of 15 (or 30 per cent) of those seen to be included in the cull, perhaps allowing a few more in recognition of those unseen and unrecorded in spring. In fact, contrary to much early teaching, most of the major territories in the forest remain occupied for the entire year, and it is possible to allot a spring territory to a particular male from as early as February. This is not the case with the more open agricultural areas, where bucks often appear during

The census work in March is essential in gauging the total number of mature bucks present on the Estate.

The more open agricultural areas are where bucks often appear during the season to occupy and exploit territories according to their attractiveness at a particular time in the growing season.

the season to occupy and exploit territories according to their attractiveness at a particular time in the growing season.

So, having established how many you are going to shoot, how, when stalking, do you decide whether a buck that appears is to be included in the cull or not? First, there must always be a conscious decision-making process, and second, you must remember what has happened previously. In other words, is the cull progressing according to plan, or have you made so many mistakes that you cannot afford to make another? It is no good completing your target cull in numbers, and then adding another half dozen because up to now you haven't shot any really old bucks – and by the way that one over there looks old! So how do you establish age in the field? Coat change is a useful indicator, with old bucks tending to retain their winter coats right up into late May. Gait is also important – unstressed, an old buck will hold its head low and patrol its territory with quiet confidence – a younger buck will be constantly alert and looking around, with head high. Sharp white tines are associated with youth; short, blunt tines and swept-back antlers are associated with age (see Figures 4.1 and 4.2). Sloping coronets, as previously mentioned, are the most useful and reliable indicator of age, but should be taken in context with all the other indicators. Remember that a fat belly will often mean an old beast – as its teeth begin to wear, a deer will find digestion and conversion of cellulose increasingly difficult, and a starving deer will often die with a full rather than an empty belly.

Medal trophies have formed an important part of my life as a professional deer manager, and are indicative of the continuing success of our deer management input. However, it must be understood that I do not go out to shoot all the medal trophies that are seen in a particular season; rather, a buck's propriety for culling is assessed by age-class, and if it has a

Coat change is a useful indicator, with old bucks tending to retain their winter coats right up into late May.

BELOW *Sharp white tines are associated with youth.*

good head, then so be it. Thus medal-class trophies are represented in the cull at the same percentage as they exist in the census. Again, medal heads often take us by surprise, tall trophies, which at 100m look outstanding, often suggest youth and thus are liable to fall disappointingly below medal class. Short thick heads often deceive in the other direction, and give a pleasant surprise.

While the doe cull represents an arduous challenge, the buck cull is a relatively relaxed affair. The season and the days are long, and the weather much more accommodating. Although many stalkers are keen to start on 1st April, I believe that there are some popular misconceptions surrounding this decision. There are several factors that influence stalking success: temperature, air pressure, moon-phase, wind, rain, availability of feed, development of undergrowth, and relative territorial activity amongst the bucks. In April new grazing becomes available but it is often cold, the wind bites, and territorial behaviour becomes subdued. Stalking bucks in those circumstances can be a depressing affair. In May, particularly late May, the average temperature is higher, browse becomes plentiful, and territorial behaviour intensifies. Stalker and quarry, however, operate on completely different temporal, spatial, and sensory planes. The temporal aspect of stalking which is crucial to one's chances of coming across the right buck, at just the right time and in just the right place to permit a safe shot, is limiting enough. Add in the deer's quite different use of its senses, and its different understanding of topography (where, for example, thick bramble bushes do not represent a barrier), then it is perhaps

In May, particularly late May, the average temperature is higher, browse becomes plentiful, and territorial behaviour intensifies.

OPPOSITE PAGE *Medal class trophies are represented in the cull at the same percentage as they exist in the census.* Photo by Chris Howard

extraordinary that stalker and quarry ever come together. Therefore, to increase the chances of colliding with one's buck, it helps if he is moving around. Thus in late May, the bucks are at their most active outside the rut, and are readily stalked despite the growth in ground cover. What I am saying is that I have spent many miserable days stalking in April trying to 'beat the cover', but enjoyed unprecedented success in late May in spite of it. Furthermore, it just does not seem right to be stalking a buck in its coat change, and with uncoloured antlers. In terms of an overall deer management plan however, the luxury of waiting might not be available to you. April can be a useful month to make significant inroads to your yearling buck cull. It can also be essential in areas of particularly dense woodland, where a cull not completed in April might not be completed at all.

June is another maligned month. We read that because of the summer growth stalking becomes impossible, but this is not my experience, either in terms of yearlings or in terms of the very old bucks. The only time that I avoid is early July, when everything seems to go quiet before the rut.

The rut can be both the most exciting and the most frustrating time to stalk. From about 20th July, intense activity starts, but just when things seem unstoppable the weather sometimes gets cooler, and activity ceases just as suddenly. 'Calling' can never be considered to be unfair, as in reality its success is no more guaranteed than any other form of approach. For years I struggled, calling the occasional buck and frightening many, never feeling confident with either the noises or the technique that I was using. However, I was then lucky enough to meet two important influences in teaching this really fascinating art. First the late Prinz Heinrich Reuss taught me the *technique*, and more latterly Bertram Quadt (grandson of the famous author, the Duke Albrecht of Bavaria), using his grandfather's original calls, taught me much about the actual *tone* of calling. I am no expert today, but enjoy some success, importantly with the self-confidence that at least I am doing more or less the right thing.

Of the two, technique and tone, technique is probably the more important. It is no good just going into the woods and blowing – you must plan carefully for wind and select a spot where the deer can be seen approaching, but with sufficient cover to ensure that the buck is confident to make the approach in reaction to your call. Stalk into your chosen site and settle down, ensuring that you have a clear view behind you as well as in the direction from which you expect the buck to come. Then allow several minutes to accustom yourself to the environment. Whichever call you use, start with a quiet and plaintive series of 'pheeps', then wait several minutes to assess any reaction. Repeat with a louder and perhaps deeper series of calls, and wait again. The great excitement from calling is in 'springing' a buck, when it charges in almost immediately in response to the call. I am aware that most literature suggest waiting at least an hour before moving to the next calling site, but my own preference is to try to 'spring' the buck, and if this is unsuccessful I move on within 15 minutes

The great excitement from calling is in 'springing' a buck, when he charges in almost immediately in response to the call.

or so. No doubt I am missing many opportunities, but my feeling is that if one really waits long enough a buck is bound to come anyway, and this does not constitute the excitement of the rut.

Success is by no means guaranteed: weather conditions are a major factor – with wind, cold or low pressure significantly reducing one's chances. There is also a 'crucial' period which will vary from day to day – it might be first thing in the morning on one day, but 4.00pm on another, and it is almost certain to coincide with your chosen break for breakfast or lunch! More often than not, a doe will come first, maybe followed by a buck, but maybe not. The excitement is intense, and even if and when the buck does come, the chance of a shot may not present itself.

The rut is also a great time to photograph deer, not least because increased daytime activity amongst the deer means that the camera is not always crying out for more light. That elusive picture of mating roe still remains an objective.

After the rut, things definitely go quiet for a few weeks, although the yearlings will take the opportunity to range free from the territorial aggression of the now-exhausted mature bucks. If the cull is still not complete, then there will be ample opportunities on the stubble fields

BELOW AND OPPOSITE PAGE
The rut is also a great time to photograph deer.

from late August through September to select the final few. There are also nearly always a few injured mature bucks that will be found dead or dying. The most common source of injury is through sparring head to head, when small abrasions may be caused to the frontal bone just beneath the coronets. These will often become infected and, because the deer cannot lick the wound, will attract fly strike. Eggs will be laid, and once the maggot has hatched, the deer will suffer a slow and obviously distressing death as the maggots eventually eat right through the flesh and skull into the brain. Any buck seen in late summer shaking its head, and appearing

The result of fly strike; any buck thus affected should be put out of its misery.

unwary, should be suspected of having such an injury, and put out of its misery immediately.

My personal preference is to complete the buck cull by the end of September, giving them a clear month, as in March, with no stalking pressure before the start of the doe cull in November. This serves as yet another way of limiting stress and ensuring visible, high-quality deer.

In summary therefore, although counting is impossible, ageing in the field is impossible, and mistakes are bound to occur, it is still possible to take a structured and deliberate cull which broadly targets the very young and the very old, whilst sparing the middle-aged. In conclusion, whilst best practice is difficult to achieve, poor practice is readily avoidable.

5.

STALKING EFFICIENCY

THERE IS NO DOUBT that, as time goes on, we all become much better stalkers. This is because we come to know the ground, the deer, their behaviour and their feeding patterns. With roe deer, a recreational stalker would generally reckon on one selective cull per three outings, whilst a professional will have to work towards an average of one cull per outing. When I recorded my own success rate in 2002, my own average through November was 1.4 roe per outing. With fallow, of course, this average is dramatically reduced to the extent that it really makes little sense for a single person to attempt the cull alone. In many circumstances the only way that fallow can be *stalked* efficiently is for at least two people to work together, and even then success will rarely match that experienced with roe. In pure economic terms therefore, stalking fallow deer is not a viable exercise in terms of the costs against the returns from venison.

As our personal skills improve, the most significant advance comes in 'timing'. Live quarry is very different from a paper target in that it is three-dimensional and mobile. I have met few stalkers who really understand how to time a shot to best advantage. Much has to do with being predictive, and understanding the techniques used by the individual species to feed. Muntjac, for example, tend to dart around the forest, pausing only briefly to snatch a bite here and there. They are a small quarry, often obscured by the understorey. It is therefore essential to observe their direction of movement, and try to predict to where they will move. Then, if a clear area is seen, the rifle can be up and waiting so that if the deer passes through that area, and if it stops or can be made to stop, then

With fallow, it really makes little sense for a single person to attempt the cull alone.

there may be an opportunity for a shot. The roe, on the other hand, is a much more sedentary feeder, typically taking a few paces between each bite and stopping to look around in between. Shots should not be taken whilst its head is lowered to feed, as the vital organs are pressed into the stomach and the risk increases of the bullet striking the gullet, with resultant spillage of green matter into the chest cavity. However, when it throws the head up to look around prior to resuming its motion through the forest, this gives the perfect opportunity for a shot. Indeed the whole procedure becomes a rhythmic and predictable pattern of movement into which the chance of a shot is readily planned. If you do not understand that typical pattern, you will often misjudge your timing and lose the opportunity of a perfectly placed shot. Fallow, once feeding, are quite bovine but very aware. If you are clever enough to make your approach unseen, then a well-considered shot, or series of shots, should not be a problem.

The key is in understanding the predictive skills required. A deer may be in view, but apparently unstalkable. With experience, you should be able to predict what it is likely to do, and what therefore you can do to put yourself into a position to shoot it. Very often the solution is to do nothing, but allow the deer the time to put itself into a good position. Preemptive action will often actually *reduce* your chances, as I remember when once stalking fallow bucks with an experienced but unlucky guest. We were standing in an exposed position with a shootable buck rapidly approaching us over a field. It had not seen us, and was clearly going to walk within easy shot of us. All we had to do was to get the rifle up onto the sticks and wait motionless for the opportunity. The guest, however,

Muntjac tend to dart around the forest, pausing only briefly to snatch a bite here and there.

The roe, on the other hand, is a much more sedentary feeder, typically taking a few paces between each bite and stopping to look around in between.

chose to sink to the ground, presumably with the intention of a prone shot. The deer fortunately did not see, but continued towards us quite unaware, eventually passing just 20 metres from us, pausing briefly to inspect us, and at all times completely unshootable because of the lack of safe background from the prone position. While misjudgements of this kind demonstrate mainly lack of knowledge of the ground, they also demonstrate the importance of allowing the deer to come to you, and of selecting the most suitable shooting position for the circumstances. This is without doubt the single most important factor in stalking success or failure.

Identification of quarry can be difficult for the beginner, and perhaps the most likely time for misidentification is during the winter just after the roe bucks have cast their antlers. However there is a discernible difference in the angle of the ears (see Figure 5.1) which identifies a buck from a doe at any age, as well as the well-documented and fail-safe anal tush . The outlines of the deer's heads in Figure 5.1 are based on tracings from dozens of

If you do not understand the typical pattern, you will often misjudge your timing and lose the opportunity of a perfectly placed shot.

Allow the deer time to put itself into a good position.

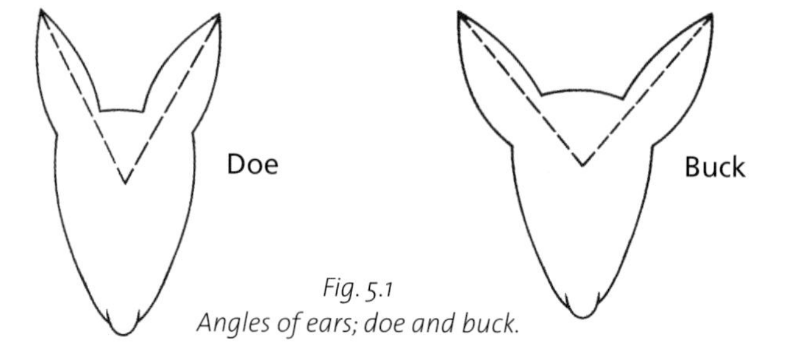

Fig. 5.1
Angles of ears; doe and buck.

photographs, and seem to show a difference of angle amounting to some 15 degrees.

One of the other factors which, as a professional, becomes increasingly important is the most effective use of daylight. Many of us, often under pressure from clients, force ourselves into the illogical situation of entering the forest before dawn. In fact, the records prove that this is a poor use of time, and often reduces success through filling the forest with unnecessary human scent before the deer are even visible. Furthermore, having repeatedly risen too early, you are then tired after two or three hours and inclined to return for breakfast – at precisely the time when the records show the roe to be at their most active. No woodland stalker can maintain the necessary concentration for more than three hours, and it is therefore important to make the best use of those three hours.

Of a sample of 2,000 consecutively stalked deer (i.e. excluding does culled in movements) including 1,180 does and 820 bucks, the results will come as a surprise to many (see Figure 5.2a–d). Of the bucks culled in the morning, see Figure 5.2a, 90 per cent of all bucks culled were shot between 05.15 and 08.30, with a clear peak around 07.00. Although 18 out of the 820 bucks were shot at very first light, this accounts for only 2 per cent of the total cull, and one has to ask oneself whether it was worth 'busting a gut' to get up so early for them, especially as there is every chance that some or many of them might have been shot at more normal times. Furthermore, particularly in April when cold north-easterly winds frequently blight the early signs of spring, bucks are often much more active after 08.00, and I believe that many stalkers are tiring and returning home just as the deer are beginning to become active. This suggests that if frost is predicted, it may pay to be brave, stay an extra hour or two in bed, and stalk between 08.00 and 01.00. (Surprisingly, the midday rut stalking accounted for only about 2.5 per cent of the total.)

Again, during the evening buck stalking period (Figure 5.2b), 91 per cent were shot between 19.00 and 21.30, with a clear peak around 20.00. So the morning has a three-hour effective period, but the evening only a two-and-a-half-hour peak of activity. Perhaps the most interesting bit of information, which surprised me as the counting progressed, is that the evening:morning success ratio is almost precisely even, at 51 per cent to 49 per cent.

Turning to the morning doe cull (Figure 5.2c), 87 per cent of all does culled were shot between 06.30 and 09.00, with a clear peak around 08.00. The later morning, however, is significantly more productive than with the bucks, and this has much to do with the feeding activity of roe does in late winter, when in certain conditions they can be active at all times of the day. Note also that people always talk of restricted winter stalking time, but here is the proof that the effective two-and-a-half-hour period between 06.30 and 09.00 is a full hour less than the effective summer buck stalking time. In the evening, however (Figure 5.2d), there is

Stalking times

Morning buck culling times (a)	
0445	3
0500	9
0515	**18**
0530	**21**
0545	**21**
0600	**31**
0615	**19**
0630	**31**
0645	**26**
0700	**60**
0715	**25**
0730	**39**
0745	**17**
0800	**31**
0815	**11**
0830	**11**
0845	0
0900	3
0915	2
0930	5
0945	0
1000	3
1015	0
1030	4
1045	1
1100	3
1115	2
1130	4
1145	1
1200	1
Total	**401 (49%)**

Evening buck culling times (b)	
1215	0
1230	1
1245	1
1300	0
1315	0
1330	0
1345	0
1400	0
1415	0
1430	1
1445	01
1500	0
1515	0
1530	0
1545	0
1600	3
1615	0
1630	2
1645	1
1700	3
1715	0
1730	1
1745	4
1800	3
1815	4
1830	7
1845	4
1900	**16**
1915	**13**
1930	**28**
1945	**37**
2000	**76**
2015	**43**
2030	**44**
2045	**46**
2100	**27**
2115	**33**
2130	**17**
2145	4
Total	**419 (51%)**

Fig. 5.2 continues on p. 71

Stalking times

Morning doe culling times (c)	
0615	**1**
0630	**8**
0645	**16**
0700	**30**
0715	**33**
0730	**41**
0745	**75**
0800	**108**
0815	**74**
0830	**61**
0845	**42**
0900	**39**
0915	7
0930	8
0945	4
1000	15
1015	5
1030	4
1045	1
1100	14
1115	0
1130	8
1145	2
1200	8
Total	**604 (51%)**

Evening doe culling times (d)	
1215	3
1230	5
1245	1
1300	1
1315	0
1330	1
1345	0
1400	6
1415	4
1430	5
1445	1
1500	**23**
1515	**11**
1530	**27**
1545	**40**
1600	**68**
1615	**72**
1630	**76**
1645	**71**
1700	**73**
1715	**33**
1730	**22**
1745	**25**
1800	8
Total	**576 (49%)**

Fig. 5.2 continued

some compensation in that the effective stalking period is slightly more extended and more evenly distributed than for the bucks, lasting two-and-three-quarter hours between 15.00 and 17.45 during, which 94 per cent of all does were shot, and with no clear peak hour.

The wise use of time, and the most efficient use of sleep, will therefore ensure the most effective and successful stalking.

One might equally asked why anyone would ever stalk alone when collaborative culling is so very much more productive. Most roe stalkers love the peace and solitude of the natural environment and of the stalk, and the long summer hours spent stalking bucks is a pleasure which can be enjoyed alone. The doe cull, however, may be quite a different story. More and more stalkers are realising that collaboration at one level or another has an exponential effect on their returns. Two people stalking or sitting more than doubles the net success, while ten on the same area exceeds even that. Of course the number of Rifles must reflect the nature and size of the ground, and if multiples are employed then strict safety criteria have to be satisfied. The principles of collaborative culling were covered in Chapter 3 and can be an extremely productive use of resources.

6.

RECORDS

SINCE THE FIRST edition of this book, there have been a number of significant changes which skew the subsequent records. The last year for us on one major Estate was 2006, and on another my principal's tenancy ran out in 2008. We have since taken a new area upon which we are now starting the long cycle of improvement.

Nevertheless few management theories anywhere in the UK or Europe are supported by data that can sustain them. The minimum period over which high-quality records may be scrutinised must be the average life-span of the species involved, which in the case of roe can be taken to be ten years. This is not to say that some roe cannot outlive this span, but there are so few roe within a population greater than ten years of age that they are statistically insignificant. As my records develop over the years, the annual analysis becomes evermore significant, and the results of each year are awaited with some apprehension – but also with increasing confidence as, each year, the statistics appear to move in the right direction. Even after five years, the management philosophy appeared to be working, and after ten years I was beginning to feel assured. Nevertheless, there was always a chance that things might show signs of deteriorating, and so I bided my time until there appeared to be certainty. Now, after 25 years, I believe I can really say that it is indeed possible to manage roe in a way that improves both their manageability and their quality. It requires a plan, self-discipline, a long-term view and the eye to age live deer. One also needs time; time to get to know one's deer and to learn which to leave and which to cull.

Of course, the first factor is to establish the physical indicators of

Bodyweight is the single best recordable indicator of health within deer populations, with average annual antler growth being an excellent test of overall quality.

success and failure, and then ensure that the records kept reflect those criteria. It is generally accepted that bodyweight is the single best recordable indicator of health within deer populations, with average annual antler growth being an excellent test of overall quality, albeit with the influence of seasonal variation. In roe, the only sensible way of assessing trophy quality is through skull weight, rather than arbitrary factors such as length, number of tines or regularity. The percentage of bucks achieving medal class is obviously irrelevant in marginal areas, and may even be statistically insignificant in the best areas of central southern England. It is, nevertheless, an interesting and important test of quality, which supports the benchmark of average trophy weight.

Equally important is to establish starting data, and decide how to measure success or, indeed, failure. When I started managing deer in southern England, it was really at the peak of their ascension and colonisation. There was relatively little deer management going on (although a fair bit of stalking) and numbers were perhaps not yet at saturation density. In other words population density was probably at its optimum for quality, and stress levels were low. Visible quality therefore was already very high, indeed the highest in Europe, and one had to decide what was

Quality was already very high.

a realistic test of success. With bodyweight and antler growth already significantly in excess of anything experienced elsewhere, it was unrealistic to assume that any improvement was possible. The conditions at the time were almost certainly as good as they could ever be, and we had therefore to assume that success was measurable simply in one's ability to maintain that quality. In the years between, many other deer management enterprises have failed to maintain that quality, either through over-exploitation of the males, or through under-exploitation of the population as a whole. My own records, however, appear to demonstrate that adherence to the deer management principles described earlier will ensure continuing high standards which, because of our excellent habitat, are unsurpassed elsewhere in Europe.

Assessing consolidated records

The consolidated records need some explanation. From 1986 the ground under management expanded to its optimum around 1991, and the overall cull thus rose over that period. The figures relate to a number of similar Estates in the same area, dominated by three major almost neighbouring landholdings of about 1,600 hectares (4,000 acres) each. The significant rise in the overall cull in the late 1990s reflects a population recovery on one of those areas, which we addressed from around 1991. The lack of full doe weight records in recent years is a failing for which I must hold myself responsible. The fact remains that, after so many years showing no significant variation, there seemed little point in continuing to analyse such records. That is not to say that we do not still record and monitor either doe carcass weights, or indeed the presence of a proportion of large does, but the evidence is clearly visible in the deer population, and the recording is done mentally. The intention now is to sample the population formally from time to time.

Buck/doe cull ratio *Figure 6.1*

This graph demonstrates how the ratio of bucks:does culled has varied over the years in response to our spring census, but has averaged at 33 per cent male and 67 per cent female. As explained previously, this policy has been developed simply to counter a perceived general over-exploitation of mature males within the county as a whole, and to work towards a 1:1

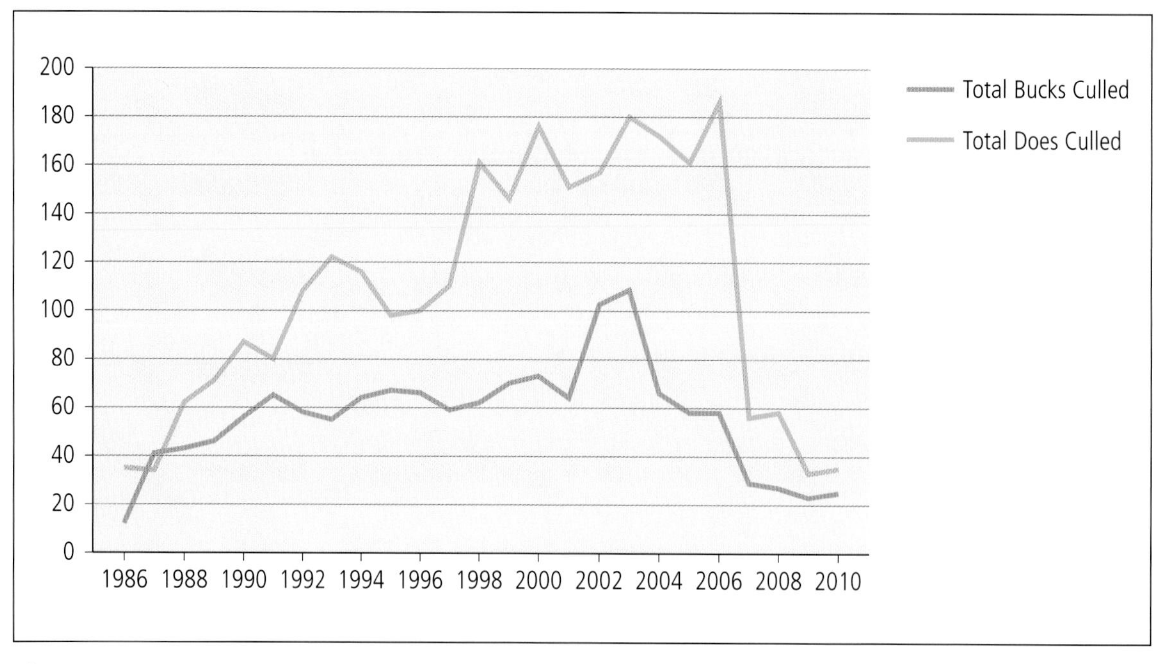

Fig. 6.1

observed spring census sex ratio. The way in which it has been achieved is by simply sparing a few extra young middle-aged bucks, which technically might have fallen within the cull. This has not resulted in increased tree damage, and one can only assume that these extra bucks have been exported from the area, either permanently or temporarily. What must be remembered is that over the county as a whole, a buck can only be shot once!

Adult buck carcass weights

Figure 6.2 (gralloched, head and feet removed)

When we started, the deer were in excellent condition and mature buck carcasses, at an average 18.4kg clean, were staggeringly high. Although in Scotland occasional very large carcasses are recorded (in excess of 27kg clean), this is very rare in central southern England. Indeed, from some 1,000 buck carcasses recorded, only two of 23kg have been noted. More typical for us is several large bucks per annum of about 20.5kg each, but even this represents only a 15 per cent weight premium over a 15-year average. I must say that I never expected average carcass weights to remain so steady, with a recorded annual variation range of only 0.9kg – just a 5 per cent deviation from the average, much of which is almost certainly to do with the weather conditions in late winter and early spring.

When we started, the deer were in excellent condition and mature buck carcasses, at an average 40.5 lb (18.4 kg) clean, were staggeringly high.
Photo by Brian Phipps

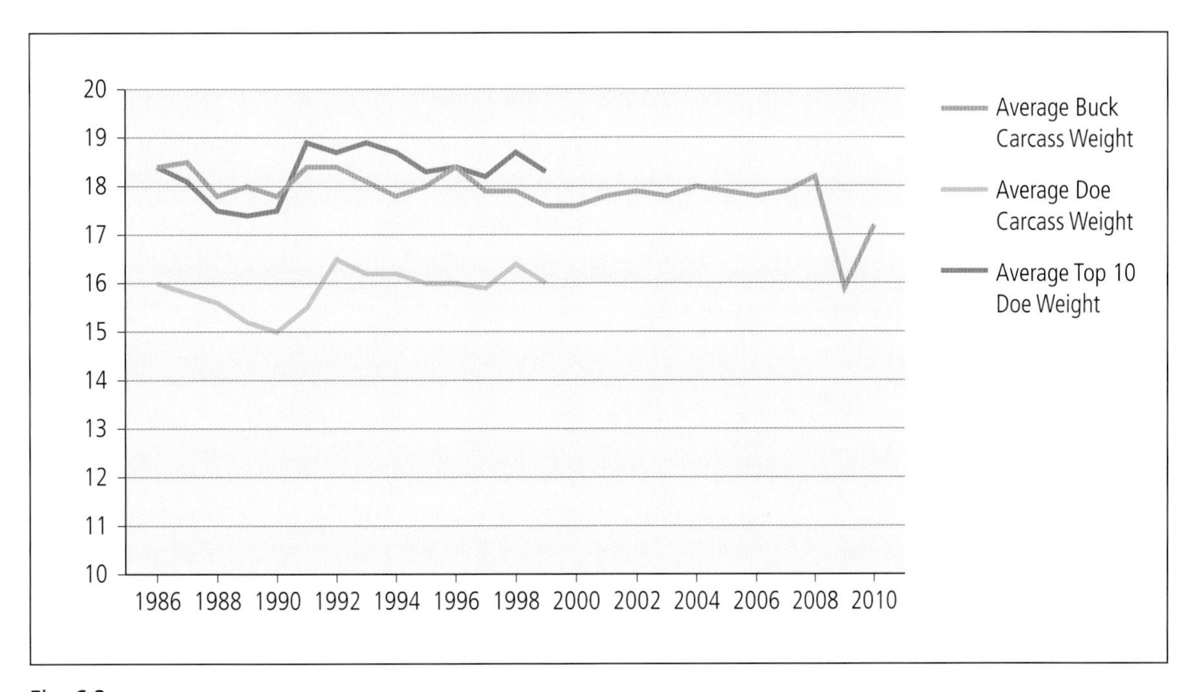

Fig. 6.2

Adult doe carcass weights

Figure 6.2 (gralloched, head and feet removed)

A similar story emerged with the doe carcasses, with a range of only 1.5kg, just under a 10 per cent deviation from the annual average, over 12 years. Again, it has really surprised me just how stable these weights have remained, and of course they are considerably greater than those previously recorded around Europe. With the total sample being so much greater among the does, we also recorded the average of the top ten carcass weights. This mirrored the average of all weights, albeit at the higher level of 18.3kg, but within a similar range of 1.4kg, or 7.5 per cent, deviation from the average. These are really very big does, and again the total number taken each year is steady at about seven per year and represents only about 5 per cent of the total number of does culled.

Trophy weights *Figure 6.3 (grams, standard cut)*

The recorded trophy weights of mature culled bucks represent the highest-known quality. Any deficit of management policy here would surely have shown itself over the years in decreasing average antler growth. While quality has varied from year to year depending mainly upon weather conditions, with occasional very poor years, the overall picture remains one of the maintenance of high quality. The annual deviation from the average amounts to no more than 27g – a mere 7.5 per cent of the annual average – and is therefore probably insignificant. A peak occurred in 2006–2008 and this must be significant in that it has come towards the end of the period of recording, and therefore shows that the policy continues to succeed.

Medal trophies *Figure 6.4*

It is rather difficult to analyse statistically the total number of medal trophies per annum. At around ten per annum, the sample number is rather low. However, taken together with physical observations during the spring census, a picture emerges of continuing high quality, and steady production of premium-quality trophies. Furthermore, as a percentage of the total number of mature bucks culled, the occurrence of medal trophies has remained constant at around 27 per cent of all mature bucks. The best year in this respect was 2006 – again, in the latter part of the period under scrutiny.

Summary

Looking at all of the graphs together, a clear high point emerges in terms of quality and stability around 1991–1993, and again from 2003–2008. This is interesting because it supports some empirical evidence provided

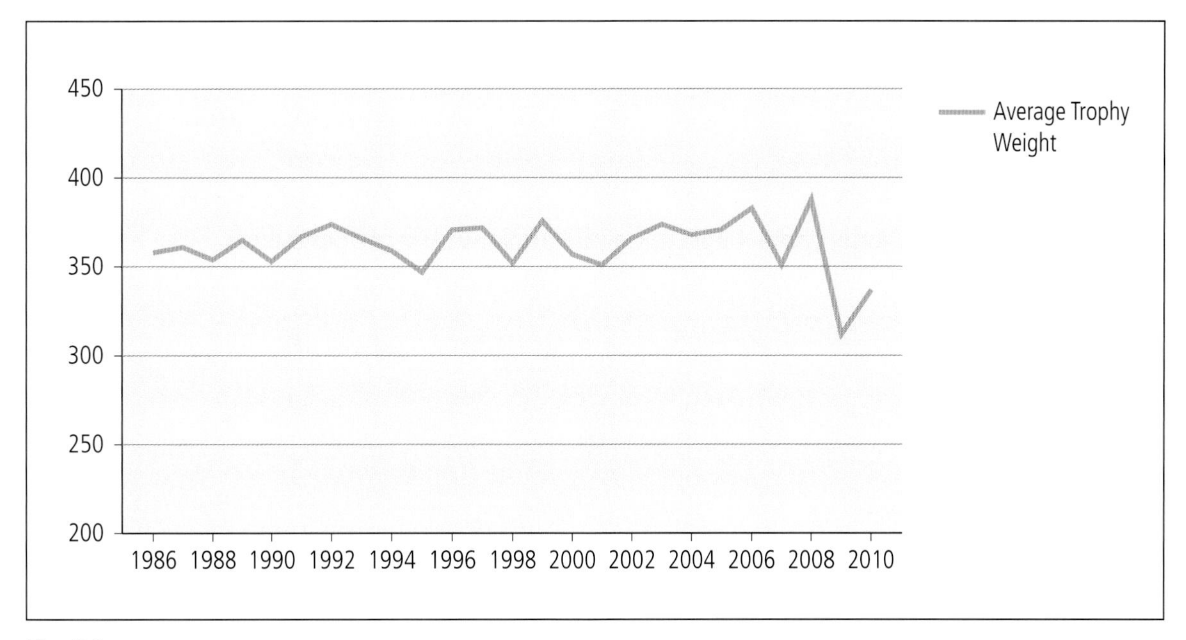

Fig. 6.3

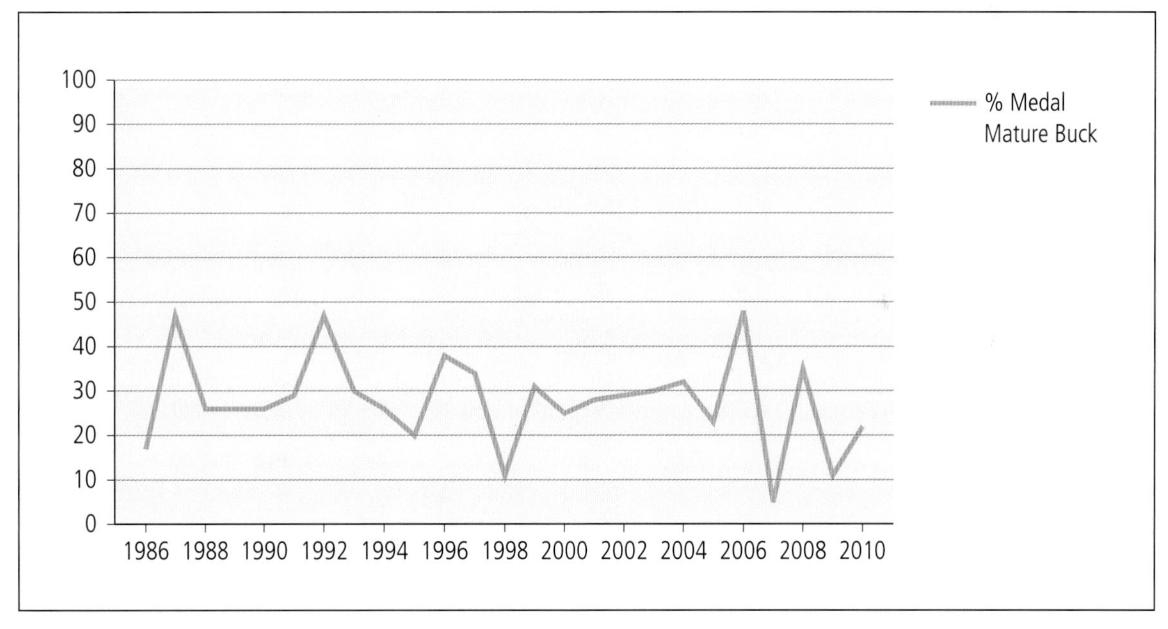

Fig. 6.4

by professionals that relatively good summers followed by relatively dry, cold winters appear to provide the best conditions in which roe may thrive. Some of the strongest and some of the oddest trophies that have been taken over the years appear in Appendices 1 and 2 respectively at the end of the book.

Table 6.1 Consolidated records 1986–2010

Roe deer

	1986	1987	1988	1989	1990	1991	1992	1993	1994
Av. Trophy Weight	358	361	354	365	353	367	374	366	359
Total Medals	2	8	6	6	7	10	15	11	11
Bronze/Silver/Gold	1/1/0	5/3/1	2/4/0	4/0/2	5/1/1	5/3/2	8/5/2	7/1/3	6/3/2
% Medal Mature Buck	17	47	26	26	26	29	47	30	26
Av. Buck Carcass Wt	18.4	18.5	17.8	18.0	17.8	18.4	18.4	18.1	17.8
Av. Doe Carcass Wt	16.0	15.8	15.6	15.2	15.0	15.5	16.5	16.2	16.2
Av. Top 10 Doe Wt	18.4	18.1	17.5	17.4	17.5	18.9	18.7	18.9	18.7
Nos Does + 18kg(40lb)	2	4	3	1	2	8	18	14	13
Total Bucks Culled	12	41	43	46	56	65	58	55	64
Total Does Culled	35	34	62	71	87	80	108	122	116
Buck/Doe Ratio	26:74	55:45	41:59	39:61	32:68	45:55	35:65	31:69	36:64

Table continued 1995–2003

Roe deer

	1995	1996	1997	1998	1999	2000	2001	2002	2003
Av. Trophy Weight	347	371	372	352	376	357	351	366	374
Total Medals	9	17	13	4	14	11	7	13	17
Bronze/Silver/Gold	5/3/1	7/6/4	7/6/0	2/2/0	8/3/3	8/2/1	4/3/0	8/4/1	9/6/2
% Medal Mature Buck	20	38	34	11	31	25	28	29	30
Av. Buck Carcass Wt	18.0	18.4	17.9	17.9	17.6	17.6	17.8	17.9	17.8
Av. Doe Carcass Wt	16.0	16.0	15.9	16.4	16.0				
Av. Top 10 Doe Wt	18.3	18.4	18.2	18.7	18.3				
Nos Does + 18kg(40lb)	6	4	5	9	7				
Total Bucks Culled	67	66	59	62	70	73	64	102	108
Total Does Culled	98	100	110	161	146	176	151	157	180
Buck/Doe Ratio	41:59	40:60	35:65	28:72	32:68	28:72	30:70	39:61	37:63

Table continued 2004–2010

Roe deer

	2004	2005	2006	2007	2008	2009	2010
Av. Trophy Weight	368	371	383	351	388	312	337
Total Medals	11	7	17	1	7	2	4
Bronze/Silver/Gold	3/7/1	1/3/3	11/4/2	0/0/1	2/3/2	0/1/1	3/1/0
% Medal Mature Buck	32	23	48	5	35	11	22
Av. Buck Carcass Wt	18.0	17.9	17.8	17.9	18.2	15.9	17.2
Total Bucks Culled	66	58	58	29	27	23	25
Total Does Culled	172	161	186	56	58	33	(35)
Buck/Doe Ratio	28:72	26:74	24:76	34:66	32:68	41:59	42:58

NB The period 2004–2010 represents one of many changes. 2006 was the last year on one major Estate, whilst my principal switched Estates in 2009 to commence a complete restructure and improvement of a new area.

Photo by Chris Howard

7.

CASE STUDY 1: A 15-year Experiment

IN 1993 I WAS APPROACHED by an East Hampshire Estate to undertake a survey of deer. My report led to a relationship lasting some 15 years. I would like to think that the results of what we were able to achieve and the manner in which we achieved them could well be seen as a template for successful deer management.

The Estate extended to some 1,520 hectares (3,750 acres) with over 400 hectares (1,000 acres) of woodland, including a large central block. It was well contoured, private, accessible and in mixed cropping – ideal deer country. I remember being very excited by this opportunity and in all honesty had almost written my report before the visit, believing that it would only be a matter of filling in the numbers after the event. I knew the surrounding area well and density figures for deer were fairly well established for the area. How wrong could I be!

Over a week on the Estate I counted just 12 roe and four fallow deer, none of which were seen in the fields and all of which were extremely nervous of motor vehicles. There were no adult roe bucks. It was extraordinary.

Under these circumstances I had to rip up my draft report, seek instructions from the Estate and start from scratch. At the same time I was aware that here was a fantastic opportunity to deliver on a range of issues in a measurable and visible manner. The Estate was fortunately clear in its requirements – for a healthy, visible, high-quality stock of roe and fallow deer which the owners could enjoy and stalk. We therefore devised a plan in three stages:

1. Population recovery.
2. Establishing a cull and reducing stress.
3. Recreational stalking and increasing quality.

At this moment I predicted a timescale of five years to see a full population recovery and ten years to bring the Estate up to its full potential.

Population recovery

Year 1 (1993)

Before planning for a rise in population it was important to establish why numbers were so very low and what might be reasonable to expect. Typical roe density for the area was some 50 per sq km of woodland, suggesting a potential population of at least 200 roe. Clearly we had some way to go! Analysis of the records held by the Estate, using any of the accepted models of population dynamics, threw little light upon this shortfall and we concluded that the low population must be a consequence of extraordinary mortality, massive poaching or unrecorded culling. We had to accept that the latter was the most likely and orders were issued to stop all culling. From previous experience my recommendations were that full recovery might take up to five years. At this stage there was little else we could do but wait until March the next year to see what happened.

Year 2 (1994)

In the second year, using the same census method at the same time of year, the number of roe deer recorded increased from 12 to 46 (with more than 12 being seen on the first night alone), and fallow deer from 4 to 19. One cannot expect to be able to record anything like all of the deer, and we worked on the assumption that we were recording something like 50 per cent of the roe present. Large woodlands make observing and counting roe more difficult and on many Estates with a more typical 15 per cent woodland cover I would expect to able to count as much as 70 per cent of the population.

So, despite still modest population density, things had improved rather more quickly than I had anticipated. It is almost certain that, given such a low population of deer in year 1, the Estate had benefited not only from a successful breeding year, but also from significant immigration. However, sex ratios of the roe were badly skewed in favour of does, leaving an observed sex ratio of 3:1 (does:bucks). Furthermore, no mature bucks were seen. Both species were occasionally seen in the fields, but remained nervous and retreated to the safety of the woods at the approach of a vehicle. Muntjac were now apparently present on the Estate, though none were recorded during the actual census. The recommendation was to continue the moratorium on culling.

Year 3 (1995)

By the third year the number of roe recorded had increased to 58, indicating a population of well over 100 but still below the expected 200+ optimum for the area. They were now spreading to all parts of the Estate and we also recorded seven adult roe bucks, which was extremely encouraging. The deer were much less nervous of vehicles, now being 'sensibly cautious' rather than 'alarmed'. Fallow numbers had also increased and five full-headed bucks were now regularly seen around the Estate. We also recorded our first muntjac. A reasonable recovery in numbers had thus occurred in as little as three years and we were looking to start some selective culling. As part of a single day's manoeuvre, 19 roe does and 12 fallow does were culled, this method limiting the stress of being stalked to a minimum.

Year 4 (1996)

Despite this cull (actually achieved in February 1996) roe numbers had again increased, now to 83 recorded during the census period, and fallow numbers had risen to 38 recorded. Additionally, 12 adult roe bucks had taken up territories and a few were now old enough to be included in a cull. We assumed that Stage 1 of our effort had now been reached exactly as predicted.

Table 7.1 Stage 1 population recovery 1993–1996

Year	Number of recorded deer			
	Roe	**Mature roe bucks**	**Fallow**	**Muntjac**
1993	12	—	4	—
1994	46	—	19	—
1995	58	7	28	1
1996	83	12	38	1

Establishing a cull: years 5–8 (1997–2000)

During this period there was no formal census undertaken, and my input was restricted to a monitoring role and to the organisation of the annual doe cull, together with some guest stalking. Cull returns showed that population density of all three species was clearly continuing to rise, along with the number of adult males. Sex ratios were also approaching parity and the deer were becoming increasingly more stable to motor vehicles. The policy of reducing stress in the population had been achieved and a marked improvement in quality was beginning to become apparent. A very few adult bucks were shot and in 1999, quite by chance,

I. Botterman, 150 points.

the first medal trophy was taken, a great gold of 150 points to a very lucky Belgian hunter. Medal-quality trophies were not to become a regular feature for some years to come. However, deer numbers overall were clearly continuing to rise and we did not feel that we were shooting sufficient fallow deer.

Table 7.2 Stage 2 Progression of cull (1996–2000)

Year	Deer culled					
	Roe		**Fallow**		**Muntjac**	
	Bucks	Does	Bucks	Does	Bucks	Does
1996	—	19	—	12	—	
1997	—	20	—	11	—	
1998	5	27	—	3	—	
1999	3	38	—	7	—	2
2000	7	34	—	5	—	2

Recreational stalking and quality: years 9–14 (2001–2006)

We were now entering the final stage of our plan. Numbers had recovered satisfactorily, the deer were steady to motor vehicles and we had established an annual cull day. The instructions were now to improve quality and continue to improve the prospects for recreational stalking on the Estate. In effect this meant more good bucks for the owner and his guests to stalk. Up to this stage the cull of bucks had been insignificant in numbers and 'predatory', and of no particular impact on the overall management. From now on I would be engaged to ensure that the buck cull targeted only the poor young and very old, leaving the middle-aged bucks to improve. The objective was to effect a measurable improvement in antler quality.

A formal spring census was put back in place, but in addition to simply counting the deer I would also identify and record as many as possible of the territorial adult bucks. Those considered old enough to be included in the cull would be highlighted and these would be the only bucks targeted during the summer. Where possible the buck would be described and even photographed, giving whoever was stalking the best opportunity to avoid mistakes.

2001

The 2001 census demonstrated that, despite our ever-rising cull, numbers had appeared to continue to rise (not least because the 2001 cull had been curtailed over much of the UK because of Foot and Mouth Disease). At the last formal survey in 1996 the number of roe recorded had been 83. This figure had now risen to 114. This, in itself, was very satisfactory, representing no more than the average density for the area but, in my opinion, the optimum for the ground and a density which we should not attempt to exceed. Of greater significance was the number of mature territorial bucks now present on the ground. Back in 1996, 12 had been recorded. This number had now increased to 32, suggesting a reasonable surplus available to stalk.

Fallow numbers had also risen significantly, with several formidable herds ranging over the Estate. I was no longer able to count them with any degree of accuracy; all I could say was that there were now well over 50. Muntjac sightings had risen from 'occasional' to 'regular'.

The cull recommendation therefore rose significantly, and the cull for the year ended at 20 roe bucks (about 50:50 young:old, including one bronze medal buck), 65 roe does, nine fallow bucks and 23 fallow does. Additionally and surprisingly, five muntjac bucks and seven muntjac does also ended up in the larder. The feeling was that we had found the right level of roe doe cull, but that there was room to increase the buck cull as the situation improved. With regard to the fallow we felt that we were well on the way to taking effective control of the herds, and we had no doubt that the surprising cull of muntjac would mean we had seen the last of them. As things turned out, we were pretty well spot-on with the roe, but it would take further efforts to stabilise the fallow and we were to discover that muntjac were now definitely here to stay!

2002–2006

During these years the formal spring census continued and we now included a formal autumn report as well. Infrastructure was improved dramatically, with the construction of a purpose-built larder, incinerator and approved rifle range. A register of high seats was maintained and a system of inspection put in place. We also experimented with the KNZ mineral licks, and I remain convinced that they have some measurable benefit upon roe quality.

The number of roe remained stable, but with careful selection of the bucks the sex ratio improved and the total number increased as a proportion of the population. Quality also improved, as measured by the number of medal-class bucks included in the cull and, of course, in average trophy quality which we started to record in 2003. The number of fallow continued to rise despite a significant cull in 2002, and the number of muntjac remained stable despite a rising cull.

The increasing number of adult roe bucks available for stalking was a key indicator that we were successfully targeting the older bucks. The rising quality further confirmed this success, and represented the culmination of a long-term plan. I remember telling the owners in 2005 that there was little more I could do for them – quality was now as high as anywhere in England and, in eight years, two years ahead of prediction, we had achieved everything that I had promised. It was therefore particularly gratifying that in 2006, my last year, we equalled every record of any Estate I have ever managed with an average annual trophy weight (standard cut and dry) of 385g *and* from a cull of 15 mature roe bucks, seven (47 per cent) achieving medal status, two gold and five bronze. The total cull of 29 bucks included 14 yearlings.

If there was one shortcoming it was in our failure to reduce the fallow numbers, but set against this was the satisfaction that we now had a reasonable herd of full-headed palmate bucks resident on the Estate.

Table 7.3 Population census 2001–2006

Year	Number of recorded deer			
	Roe	**Mature roe bucks**	**Fallow**	**Muntjac**
2001	114	32	50+	Regular
2002	151	44	100 Est.	Occasional
2003	129	49	150 Est.	Regular
2004	138	42	150 Est.	Regular
2005	130	36	150 Est.	Regular
2006	144	39	150 Est.	Regular

Table 7.4 Progression of cull (2001–2006)

Year	Deer culled							
	Roe				Fallow		Muntjac	
	Bucks	No. medals	Av. trophy wt (gm)	Does	Bucks	Does	Bucks	Does
2001	20	1B	—	65	9	23	5	7
2002	36	—	—	66	13	40	7	5
2003	36	2B, 1S	349	66	12	19	8	12
2004	32	2B, 1S	360	58	6	27	15	19
2005	26	3S, 1G	375	47	7	20	11	9
2006	29	5B, 2G	385	72	5	14	15	12

ON THIS PAGE AND OVERLEAF
A sample of roe bucks taken in the latter part of the period of management.

I was extremely fortunate to have had this opportunity, not least because it was such a broad and far-reaching project. That I was able to see it through to its conclusion means the entire lifespan of the species was covered. That I was able to deliver on every objective confirms that deer management really does work. If there are any key lessons to be learnt they are most certainly that live-ageing of deer to a reasonable degree of accuracy is paramount, as is self-discipline and the avoidance of over-culling. In my experience, where things go wrong, it is quite simply that too many bucks are being shot and, of those bucks, too many are not old enough. It takes time and experience to be able to judge age and it is only now that I fully understand the value of what a conscientious deer manager can deliver. I wish this was more widely recognised by those responsible for land management.

Table 7.5 Consolidated records – Case Study 1

Year	Deer recorded				Deer culled								
	Roe	Mature roe bucks	Fallow	Muntjac	Roe					Fallow		Muntjac	
					Bucks		No. of Medals	Av. Trophy Wt (gm)	Does	Bucks	Does	Bucks	Does
					M	Y							
1993	12		4										
1994	46		19										
1995	58	7	28	1									
1996	83	12	38	1					19		12		
1997									20		11		
1998									27		3		
1999							1G		38		7		2
2000									34		5		2
2001	114	32	47	Regular	10	10	1B		65	9	23	5	7
2002	151	44	100E	Occasional	18	18			66	13	40	7	5
2003	129	49	100E	Occasional	18	18	2B, 1S	349	66	12	19	8	12
2004	138	42	100E	Occasional	15	17	2B, 1S	360	58	6	27	15	19
2005	130	36	100E	Occasional	13	13	3S, 1G	375	47	7	20	11	9
2006	144	39	100E	Occasional	15	14	5B, 2G	385	72	5	14	15	12

8.

CASE STUDY 2: 10 Years of Selective Culling

THIS CASE STUDY represents a very different scenario from that of the previous chapter. This was another Hampshire Estate of some 1,620 hectares (4,000 acres) with over 400 hectares (1,000 acres) of woodland. Well contoured with excellent access, it has beautiful ancient semi-natural woodland, and a history of good management. Containing roe, fallow and muntjac it is without doubt the finest stalking area in the south of England. My involvement extended over about 15 years, and between 1999 and 2008 I was the stalking tenant's professional stalker. Our aim was to improve quality and the method was a ruthless adherence to a policy of shooting only older bucks. I am not saying that we never made mistakes, indeed 10 per cent of the bucks culled probably turned out to be younger than we had expected or intended. But nine times out of ten we were right and, although ten years is not that long as a measurement of management, I believe the records demonstrate that we were successful in our objectives.

In our first few years, the forest was full of old bucks which had retreated into the woodland and never ventured outside. The middle-aged bucks occupied the better territories of the forest edge and were of good quality, but the older bucks were weak in antler and quite clearly spare. We set about these and not surprisingly ended up with a low average trophy weight of 344g and just 15 per cent of the mature bucks achieved medal class. This was below the area average and even the medal levels were low, with two bronzes and a silver.

Year 2 (2000) yielded immediate results. Although we continued to cull weak older bucks in the forest, there were fewer of these as a percentage of the population as a whole. Trophy quality increased, to an average of 363g, and 20 per cent of the mature bucks achieved medal class. Of the four medals there were three bronzes and a silver.

Year 3 (2001) seemed to be extremely encouraging, with 30 per cent of the mature bucks now achieving medal class, and the medal levels were improving with three silvers and one bronze. Average trophy weight, however, was marginally down, albeit at a respectable 352g, and statistically insignificant.

Year 4 (2002) was a strange year in that there was an enormous cohort intake of yearling bucks. This meant that we had to make a significant increase in the yearling cull, amounting to double the average over the ten years. Whether this influx had any effect on the major bucks is uncertain, but the percentage achieving medal class slipped back to 20 per cent and the quality of medals was back to three bronzes and one silver. Average trophy weight remained good at 358g.

We were now halfway through our tenure and if our objectives were to be met it should have become evident by now. Not only did there seem to be an awful lot of mature bucks about, but they were also of excellent quality. That year (2003) we culled an additional five mature bucks and 32 per cent of the mature bucks achieved medal class, which meant a leap from four medals a year for the past three years to eight medals: four bronzes, three silvers and the *first gold*. Average trophy weight returned to 362g and we felt that we had made real progress.

Year 6 (2004) continued this progress and in all probability I now felt that we could go no further. With an average trophy weight of 377g, 44 per cent of all mature bucks achieving medal class and a total of eight medals to include a bronze, six silvers and a gold, there were few areas which could boast better quality.

Year 7 (2005) was a strange year. Statistically, it was apparently poorer, with only 16 per cent of bucks achieving medal class, an average trophy weight of 367g and just three medals. However, the medals were a bronze and two wonderful golds. Gold medals are a rare and much sought-after prize and thus there could be no disappointment.

Year 8 (2006) represented the pinnacle of achievement and demonstrated to us that the policy of constantly targeting the old bucks and sparing the middle-aged had most definitely brought the results for which we had worked and hoped. That year, ten medals were taken, representing an extraordinary 50 per cent of all mature bucks shot. The tally was made up of six bronzes and four silvers, and average trophy weight had risen to an almost unbeatable 382g.

Sadly, nature has a habit of reminding us that we are not completely in control. Just occasionally we have a bad year, partly owing to the winter weather and partly to the mast and acorn harvest. No one can be quite sure what makes a great year and what makes a stinker, but Year 9 (2007) was a stinker! Of the 21 mature bucks shot, only one (5 per cent) was of medal class – but at least it was a gold medal. Average trophy weight remained at 351g and under normal circumstances this would still be seen as good. It would have been sad to have ended on such a dreadful note and we approached our last year (2008) with some trepidation.

Over all the years that I had spent on that Estate, I had never seen a monster. There were plenty of excellent golds, but nothing of 150 or 160 points. It always surprised me that amongst such fantastic quality we never shot, nor even saw, one of these monsters. However, during the spring census I photographed a real contender – thick in beam, with an extra brow tine, an extra back tine and completely tame to the motor vehicle, thus giving me plenty of time to study it. It was next to the only open clearing in that part of the forest, and I assumed it must live there. However, despite putting up a temporary seat which the tenant sat in many times, we never saw this buck feeding there – although it was frequently seen and photographed sitting in exactly the same place as on the first sighting. April and May went by but this buck eluded us and I began to think it was dead. If not, then the rut would be our last opportunity ever to shoot it, but I had still not, by then, seen it for some time. We tried calling around its lying area, but to no effect. On a hunch, I wondered whether I had been completely wrong about this buck's feeding habits. Had it been feeding the 'wrong' side of its lying area where the food seemed far less good? We tried and, on the last stalking day of my principal's tenure, after an exhaustive and noisy calling session followed by a lengthy wait, it came. A great buck to end a wonderful tenure. With long, thick beams, two extra tines and great presence, there was lack of coronets and thus the volume necessary to achieve enormity – but a great gold in any event.

Of the 20 mature bucks shot that year, seven (35 per cent) achieved medal class, consisting of two bronzes, three silvers and two golds.

The buck was frequently seen and photographed sitting in exactly the same place.

PHOTOS BELOW *A great buck to end a wonderful tenure: Prince Alois of Löwenstein.*

The average trophy weight was as high as I ever recorded anywhere, at 388g.

Looking with the mind of a statistician at the ten years of data, a clear progression upwards in trophy quality is apparent. Of the seven gold medals six (86 per cent) were shot in the second half of our tenure. Of the 22 silvers 13 (59 per cent) were taken in the second half of our tenure. The average trophy weight in the first five years was 356g, and that of the second five years was 373g. Of the total number of medal class trophies – 52 – taken, 29 (56 per cent) were shot in the second half of our tenure. All this despite the one freak bad year occurring in Year 9.

One other extraordinary opportunity occurred while we were there. In about 2002 we saw a young buck with a white spot on its rump. This was either a yearling or a two-year-old and we decided then that this buck should be left for as long as possible in the hope that we could learn something about ageing roe bucks from this easily identifiable individual. Each year it took up exactly the same territory and, although we would not see it every day, it never appeared to stray more than a short distance from the grass field which represented the centre of its territory. It never grew a great trophy although it was similar each year. In our last year we decided that, being at least seven years old this buck should be included in the cull. The photographs below and opposite show that, in every aspect of ageing

BELOW AND OPPOSITE PAGE
In every aspect of life and death, Spot confounded us.

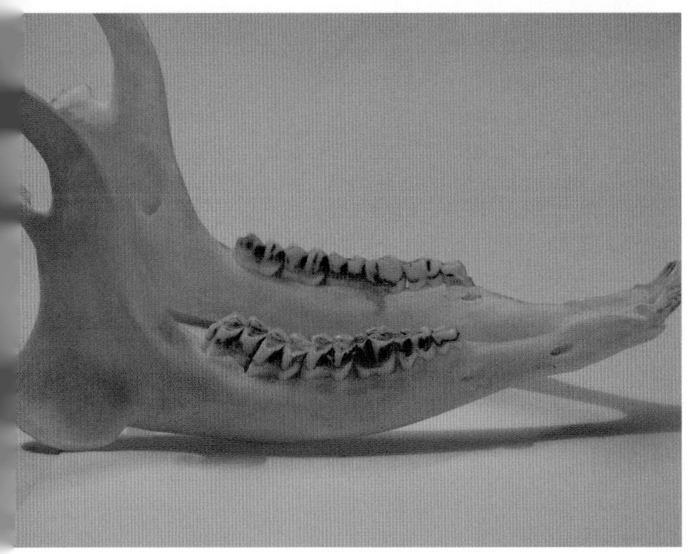

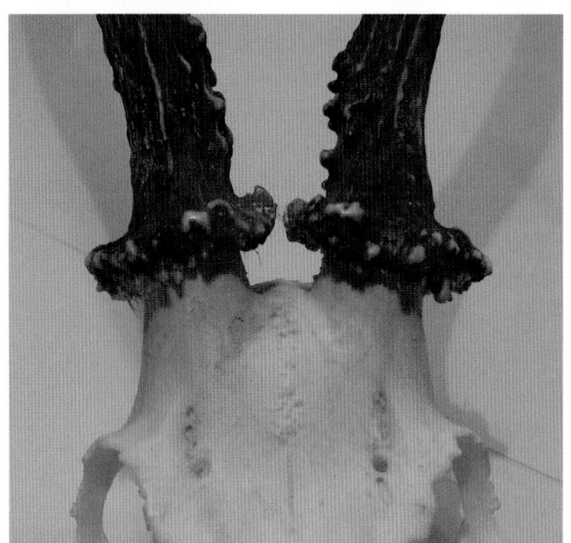

in life and death it confounded us. No particularly sloping coronets, little wear on the teeth and no very obvious grainy suture at the dividing point of the skull. If anyone had shown me the skull or teeth independently I would have aged this buck as no older that five! That said, this suggests that the majority of culled bucks which we had been ageing as 5–8 were probably actually more like 10–12 and actually exceeding the age range that we supposed existed on the Estate: gratifying indeed.

As well as the excellent roe there were also wonderful wild fallow on the Estate. No bucks at all had been shot for many years and, at last, a respectable group of males could be seen regularly. It was decided that one or two per year of the very oldest ones could be included in the cull. Some years we shot one, some years two, but one year we were lucky enough to shoot a stunning old buck which attained gold medal class. It was very old but retained thick and heavy beams and good palms. We had it set up by Colin Dunton.

A stunning old buck which attained gold medal class.

Table 8.1 Case Study 2: Consolidated cull records 1999–2008

Year	No. of bucks		Total bucks	No. of medals			Total medals	Av. trophy weight	% Mature bucks achieving medal
	Young	Mature		Bronze	Silver	Gold			
1999	7	19	26	2	1	0	3	344	15
2000	10	15	25	3	1	0	4	363	20
2001	6	14	20	1	3	0	4	352	29
2002	25	20	45	3	1	0	4	358	20
2003	18	25	43	4	3	1	8	362	32
2004	16	18	34	1	6	1	8	377	44
2005	13	19	32	1	0	2	3	367	16
2006	9	20	29	6	4	0	10	382	50
2007	7	21	28	0	0	1	1	351	5
2008	7	20	27	2	3	2	7	388	35

9.

CASE STUDY 3: Improving Commercial Potential

THIS 12-YEAR programme was a much more straightforward assignment than those mentioned previously, the brief being simply to improve the commercial value of the stalking on an Estate. It was a Hampshire Estate extending to some 1,200 hectares (3,000 acres), rather flat but well wooded and well maintained.

Prior to 1992 the deer control was still being carried out by old-fashioned deer drives with shotguns. Although we all welcomed the day when this method of control ceased, it is nevertheless interesting to note just how effective it was. In three drives per year, two in the winter for does and one in the summer for bucks, using a team of experienced keepers who never took shots beyond 20 metres, deer numbers on the Estate were kept very low. I assumed that many deer must have been wounded but the keeper assured me that this was not the case and, although they were not applying any particular science to their method, the outcome was by no means unsatisfactory. By conducting two doe drives and only one buck drive, about twice as many does were shot, and when I showed the keeper sets of antlers and asked what proportion of what type of antler was shot in the buck drives he pointed to the yearling antler as being the type most commonly shot. In other words they had been, perhaps accidently, mirroring what we teach today as best practice in cull planning, but had been utilising a method which we now acknowledge to be totally unacceptable because of its welfare considerations.

When I arrived, although there were very few deer they were of perfectly good quality and were reasonably unstressed. We set about allowing numbers to rise in a predictable way, making sure that we always retained control of the population and ensuring that quality was not compromised.

The results can be broadly seen through three four-year periods. In the first exploratory year a predatory cull of just three mature bucks was made – a very old poor specimen of just 300g and two others including a good old gold. We also took 16 does to see if this would have a visible effect on the population. It did not and, over a four-year period, a cull was established of some 20 does and 12 bucks. This cull still allowed the population to rise, but slowly and in a way that would allow us to readily address the situation if we felt it was getting out of hand.

Over the second four years the doe cull rose to 30, but the buck cull remained at 12. The long-term goal remained to increase the buck cull up to its optimum, but for the time being we were being cautious. Quality remained good.

In the final four years the doe cull reached 38 and we slowly increased the buck cull from 12 to 14 then to 20, with the final two years producing exceptional results in terms of average trophy weight – 390g and 396g respectively – and in the percentage of medals (33.3 per cent). There was no question in my mind that the Estate had reached its zenith.

However, what is interesting in this case study is what happened afterwards. The Estate had given me the project of improving the asset value and this had been reflected in a reduced annual rent. Now, with the project complete, the Estate was ready to go to the open market. They selected a European tenant who would come over for a few weeks a year. I had made it as clear as I could to the managing agents that 20 was the maximum *sustainable* cull of bucks and, of that, only ten should be mature. I also made it clear that the tenant should be compelled to take the full doe cull. Some European stalkers remain reluctant to shoot does, believing the old-fashioned and discredited dogma that does attract bucks. I was also concerned that the temptation to overshoot the bucks would be too great as we had left a healthy and very visible stock in all age-classes.

In his first year the new tenant was allowed to shoot 27 bucks, of which only a few were yearlings. Of the mature bucks most were young middle-aged. Furthermore only six does were shot. In just a few years, quality had diminished and sex ratios had become severely imbalanced.

Photo opposite by Brian Phipps

My understanding is that today the Estate struggles to find any good bucks. But this does highlight the problem with taking on tenants from a distance. If a tenant has not done the spring census themselves and learnt which are the old bucks and where they are living, then when they come over with a group of friends for a weekend it is impossible to control what is shot. The cull is bound to be biased towards the young middle-aged bucks, as the old are more wary and demand greater effort. Having shot the middle-aged, the old die anyway and there is an irreparable hole in the population if culling continues in the same fashion. This is simply proof, if it were needed, that there really is a measurable benefit from age-related selective culling.

Table 9.1 Consolidated cull records 1992–2003

Year	**Roe bucks**		**Roe does**		**Trophy details**		
	Y	M	Y	M	Skull Weights (grams, cut)	Average	Medals
1992/3	3	3	7	9	300,380,500	393	1 Gold
1993/4	7	5	12	8	280,320,365,460,580	399	1 Gold, 1 Silver
1994/5	3	6	10	11	320,340,350,385,410,415	370	2 Bronze
1995/6	9	4	13	8	300,365,375,435	369	1 Silver
1996/7	6	6	14	10	270,300,330,335,370,435	340	1 Silver
1997/8	6	6	12	12	360,360,365,365,440,450	390	2 Bronze
1998/9	6	6	20	13	300,320,325,345,370,465	352	1 Silver
1999/0	4	7	18	11	335,350,360,390,410,445,490	397	1 Bronze, 1Silver
2000/1	7	7	20	18	310,320,320,340,360,370,390	344	None
2001/2	9	6	17	13	335,365,380,380,400,420	380	2 Bronze
2002/3	11	10	21	16	290,300,360,370,400,400,400,440,450,490	390	1 Silver, 2 Bronze
2003/4	11	9	(18)	(18)	405,485,480,370,365,395,355,345,365	396	1 Gold,1 Silver, 1 Bronze

10.

FALLOW MANAGEMENT

THE MANAGEMENT OF fallow is almost certainly the most difficult of all the deer species. Few Estates have access to the entire range of their fallow, and thus even fallow *control* often presents insurmountable problems. Fallow are the species most sensitive to disturbance, and simply alter their behaviour if stalked intensively. The problems associated with fallow management are therefore twofold. First, you have no control over what happens to them when they are away from your property, and second, intensive management input at home may result in behavioural changes which temporarily obscure the management objectives. On top of that, it is sadly the case that the bucks have a much greater tendency to wander, and are thus at much greater risk of poaching or of exploitation by any neighbouring Estates with more aggressive management objectives than your own. Manipulation of quality is therefore extremely difficult to plan and often arbitrary in its success.

The effect which fallow have on the roe is widely debated, and it is certainly the case that the roe are physically disturbed by movement of fallow. In terms of quality however, it appears that the two species can live alongside each other at quite high densities, both showing very high quality. I am convinced that this is because the deer exploit the habitat in entirely different ways, the roe selectively browsing the understorey, and the fallow exploiting the arable land. Furthermore, there is clearly a 'temporal' as well as 'physical' dimension to territoriality that allows the roe unstressed enjoyment of their territories despite occasional intrusion

It is sadly the case that fallow bucks have a much greater tendency to wander, and are thus at much greater risk of a premature death.

from the fallow. Clearly, very high numbers of fallow have a generally negative influence on habitat, and with that in mind I believe that fallow and roe will only both thrive if the fallow are considered secondary to roe.

During a period of heavy poaching in the 1970s and '80s, the mature fallow bucks on one of my management areas were poached almost to extinction. The management response was to stop all culling of males, and this continued for 15 years until the population had apparently recovered. As the situation stabilised we were fortunate enough to have a magnificent and balanced population of about 150 fallow, including as many as 45 antlered males, and including medals of each category. However, taking into account just how sensitive the bucks are to outside influence, we continued to exercise extreme caution in our approach to the culling of males, and culled only one or two of the oldest ones each

In terms of quality, it appears that the roe and fallow can live alongside each other at quite high densities, both showing very high quality.

year. Furthermore, poaching has returned and if, as many predict, the economic conditions worsen then we can be sure that the pressures of the 1970s and '80s will return.

Fallow are perhaps more difficult to count than roe because of their behaviour as an often-nocturnal herding species. Direct counts at night in spring and after harvest, using night-vision equipment or even a simple lamp, can yield a useful minimum census figure. It is also the case that rutting behaviour based on the 'stand' system enables us to make apparently quite accurate estimates of their population at that time of the year, when the does and bucks congregate together in identifiable groups. However,

During a period of heavy poaching in the 1970s and '80s, mature fallow bucks were almost poached to extinction.

Rutting behaviour based on the 'stand' system enables us to make apparently quite accurate estimates of population at that time of the year.

although this gives us the necessary information on which to base the cull, a great deal more thought has to go into just how to achieve it.

Controlling fallow deer by stalking has a rapidly diminishing rate of return. They recognise a routine very quickly, and alter their habits accordingly, and I have always enjoyed much greater success by limiting stalking periods to short, intense bursts. Doubling the manpower input has the immediate effect of more than doubling the return, as two stalkers stalking the same wood will have the effect of moving the deer between them. It is in this context that the principle of 'moving' deer (see Chapter 3) is at its most effective. Taking into account the enormity of effort required by a single person to achieve the cull, and the associated disturbance to the deer, I suggest that moving is almost certainly a more efficient and sympathetic method.

On one Wiltshire Estate the fallow numbers rose enormously in cycles despite the keeper double-hatting as the stalker. Having first put in the necessary reconnaissance to establish likely patterns of movement by the fallow across the Estate, and using a system of three sitting and one stalking, four of us were able on two occasions to shoot 60 does in four days. This was very hard work from dawn until dusk, but entirely achievable and repeatable. Similarly, but on a different Estate, a single day involving 15 Rifles regularly achieves a similar result in just four hours. Yes, they learn, and you have to tweak the plan each time but, provided you put in the infrastructure, the reconnaissance and the commitment, success is not so very hard to achieve. In other words there are other options than the standard response of many government agencies of simply throwing up their hands and demanding longer seasons, night-shooting and out of season permits.

The buck cull, being necessarily small, is much easier to achieve. The prickets (yearlings with simple spikes as antlers) are active throughout their season, and the cull is readily achieved. The mature bucks, on the other hand, are elusive and even invisible for up to 11 months of the year, but become resistant to human disturbance for the one month of their rut, and are entirely predictable during that period. Indeed, although my enthusiasm for deer is principally with the roe, I look forward to the fallow rut with as much anticipation as any other single event in the deer year. The forest profile changes completely, with the disturbance caused by increased fallow activity entirely disrupting the normal territorial activity of the roe. For about a month we see very little of our numerous roe, as they appear to accede to their large and noisy 'cousins'. Scrapes and thrashing in September are the first signs of the impending rut, with groaning rarely heard before October, and with a peak of activity between about the 17th and 27th of October, depending upon the weather. Suddenly, the most elusive of deer are readily observed, and may be approached apparently without disturbance.

The ancient rutting stands remain constant every year, and in some parts of the forest there is every reason to believe that they are the same today as they have been for hundreds of years. In any event, each year the very best buck can always be found on exactly the same stand, with lesser stands holding lesser bucks. Around each stand can be found questing sorels (the first full head, often an 8-pointer but with no palmation) and sores (the first palmated head), with prickets also acting as satellites to the larger males.

The mature bucks are elusive and even invisible for 11 months of the year.

How, then, should we approach the buck cull during this period if our primary objective is the long-term management of a comparatively fragile part of the population? The first point to be recognised is that, other than during the rut, the mature bucks are rarely seen during daylight hours except when lying in ripened corn around August, or again at the very beginning of spring in early April. Although I accept that, from a purely management perspective, they can be legally shot at both times, few farmers would thank you for the damage to their crops caused by extracting a 110kg+ carcass from a field of ripe corn! Furthermore, the bucks lose an enormous amount of body condition during the rut, and are still in relatively poor condition until spring. We must therefore accept that some culling will almost certainly have to take place during the rut.

Prickets and sorels are more readily visible throughout the season, and will not therefore need to be culled heavily at this time. Since the mature bucks form such a relatively fragile part of the population, it is extremely important to be as selective as possible. The rut gives you this opportunity, as each rutting stand may be approached, observed at a safe distance, and withdrawn from, causing no apparent disturbance to the herd. An old buck can be readily identified and selected for culling with a degree of certainty impossible at other times of the year.

The important factor, as in all deer management, is planning and self-discipline. While post-rut mortality in roe is generally associated with injuries which have become infected or maggot-infested, post-rut mortality rates in fallow are primarily a result of their inability to recover body fat before winter. The roe, of course, rut in late summer and lose relatively less body condition during their rut. They then have several months in which to recover condition before times of restricted food supply. Fallow, on the other hand, complete their rut in November just as the food supply begins to be curtailed, and appear to be much more susceptible to post-rut mortality. A cull plan for fallow, therefore, should allow for much greater mortality rates amongst the mature males, and this will be reflected in a significantly limited cull within this sex and age-class. Each year on one Estate we used to find two or three of our best bucks dead about a month after the rut. With so few of them about, relative to the population as a whole, it was always particularly devastating.

Having, therefore, decided in advance how many mature bucks are to be included in the cull, restrict yourself to the minimum disturbance required to meet that cull, and avoid the temptation to use the rut as a short cut to achieving the remaining major part of the buck cull. Over-exploitation of the rutting stands will almost certainly be very short-term in its effectiveness. The reason why so few Estates can boast stable rutting stands is almost certainly a result of over-exploitation, and where there remain only prickets and sorels to sustain the rut, stands become temporary and unpredictable. I remember one Estate in Wiltshire where constant harassment of the rutting stands meant that the bucks rutted only at night and used a stand for only a few hours before moving on to another. Predictable deer management is the ultimate aim and objective of long-term planning, and destabilising your rutting stands compromises the ability to plan coherently.

Fallow management certainly depends upon great collaboration between stalkers and between Estates. It requires better infrastructure (heavier carcasses) and greater stalking skills than with roe – and also advanced marksmanship, since ranges tend to be longer than with roe and one often has to make a cull selection from a tightly packed herd milling in the middle of a field. A bit of good luck is also needed: fallow, being ranging rather than territorial, are less predictable than roe; their patterns of movement change in different weather conditions; they learn more quickly than roe and they will leave the ground if overstressed. But these are simply challenges to be overcome; it is not impossible, just hard work.

Fallow exploit the habitat in entirely different ways from other species.

11.

MUNTJAC

THE MUNTJAC IS A charming animal, and I have never understood why it attracts so much hostility. If it were proven that they had a negative influence on the native roe, either in terms of quality or indeed of simply taking territory, then perhaps I might change my mind. Accepted that they are an alien species, that they disrupt the flushing points on pheasant drives and that their small size makes them difficult to control, they are nevertheless one of our six wild species of deer and should attract the same management input as all the others. In the 25 years that I have been managing deer professionally, I have seen the muntjac arrive, increase in numbers, and then in some areas decrease again for no apparent reason. I think today that it must be the case that there are generally more in the areas I manage than there used to be, but in some of those areas they are still absent, despite being at quite high densities nearby. They are habitually held up as being destructive of bluebells, but few people are aware that while damage in some areas cannot be denied, the frequently quoted study sites are actually enclosed and may not be representative of a wild, free-living population.

Muntjac are undoubtedly habitat-specific, and do not as a first choice favour either the farm/woodland environment of Hampshire, nor its open, ancient semi-natural forest. However, I have observed that their behaviour varies enormously according to their location. In Oxfordshire, for example, before it became populated by roe, I observed muntjac behaving exactly like Hampshire roe, spending their days in cover and

The Muntjac is a charming animal, and I have never understood why it attracts so much hostility. Photo by Brian Phipps

Muntjac are undoubtedly habitat-specific.
Photo by Chris Howard

emerging from the woods at dusk to feed on the surrounding agricultural crops. This is something very rarely seen in my home area, where the muntjac seldom seem to leave the confines of the wood. What appears to be the case is that, where undergrowth is dense, then the muntjac will thrive, but not where fallow and roe are already established and dominant.

Muntjac management principles must surely mirror those of the roe. As non-seasonal breeders producing a single fawn every seven months, their breeding success must be calculated as similar to the roe – although I have no doubt that survival amongst winter-born kids must be lower and the effect of predation by foxes higher. Nevertheless their net recruitment must be on a par with roe, and the management principles must therefore be similar. Effective stalking, however, is very different. Difficult to observe, difficult to identify and requiring an acquired skill to shoot, the muntjac presents a challenge for the future.

Taking a balanced cull is difficult even with the best of intentions, and using a shoot on sight policy we invariably end up shooting more males than females and thus know – in theory but not, at the moment in practice – that we must be fighting a gently losing battle. In the small numbers which we support, control is readily undertaken during their very active period between March and April. Where density levels are high (recorded as high as 150 per sq km of woodland), it is a very different challenge. Movement is not an option, and they are very unsteady when disturbed,

Difficult to observe, difficult to identify and requiring an acquired skill to shoot, the muntjac presents a challenge for the future.

and unlikely to give an opportunity for a shot. Simply using multiple Rifles occupying high seats can bring enormous success, but I continue to wonder whether more drastic options will have to be considered in the future.

The key point, in acquiring the skills necessary to shoot muntjac, is to recognise the difference in its feeding patterns from any of the other species. Whereas the roe grazes along gently, neck down for a while

ABOVE *Using a shoot on sight policy, we invariably end up shooting more males than females.*
Photo by Chris Howard

Muntjac control is readily undertaken during their very active period between March and April.
Photo by Chris Howard

before throwing up its head to scan for potential predators, the muntjac darts between feeding areas, rarely spending much time at any one plant. This means that one's skills must be predictive, recognising where the quarry might stop and where you might make a safe shot if properly prepared in advance. It is all a matter of timing; different timing from that of the roe, but readily learnable.

In the meantime, like the roe, muntjac are territorial – which means that, provided you put in the work, a cull is usually achievable. Financially, of course, at a maximum of £15 a carcass at the time of writing, they are uneconomical to manage when compared to any of the other species – and even trophy fees are largely unsustainable as most stalkers wish simply to take one representative trophy of the species. They do, however, taste very good and of the dozen or so that I shoot each year, none get further than my own larder.

I am acutely aware that muntjac have only been present on my management areas for a relatively short time, but things have not changed much in ten years and they are still inexplicably absent on some of them. I have not yet had to eat my words since the first edition of this book, and am beginning to think that we have all been taken in by the scaremongering of the alien species stalwarts. I had a wonderful pet muntjac for a few months, brought to me as orphaned; it shared the bed with my teckel and did its business on the lino of my office floor. It climbed the stairs to sleep under my daughter's bed, but could not apparently go downstairs, having to be carried, to which it objected. One day, at about six months old, it sniffed the air by the back door and went, which was lucky as it was a male!

12.

THE HISTORY OF TROPHY MEASUREMENT IN THE UK

THE COMPARISON OF deer antlers has gone on ever since deer have been hunted for sport. In time, demand grew for rules of measurement, and numerous systems evolved, some current and some redundant, but including Rowland Ward, Douglas, Nadler, SCI (Safari Club International), Boone and Crockett and CIC (*Conseil Internationale de la Chasse*, which translates as International Council for Game and Wildlife Conservation). The Europeans embraced the CIC system, and it is now the most widely used for European deer species.

A trophy locks away many personal memories. Whether strong, weak or bizarre, beauty is in the eye of the beholder and comparisons may be invidious. Nevertheless, the average amount of antler growth in deer is one of the few reliable indicators of both annual variation and overall trend in quality. Records, particularly when cross-referenced to age and bodyweight, can demonstrate good or poor management practices, and are an invaluable management tool. It is not possible therefore to talk about qualitative management, without some reference to the historical measurement of trophies in the UK.

The CIC formulae are the most widely used measurement criteria and were devised in Europe between and after the wars. For many years the only way of getting a trophy officially measured and recorded under this system was by submission to one of the international trophy exhibitions, where each of thousands of trophies was adjudged by a panel of experts from the International Trophy Commission of the CIC. Meanwhile, in the UK, occasional lists of outstanding (by no particular definition) trophies were published by, for example, J.G. Millais and Frank Wallace. Edgar Barclay who started the annual *Shooting Times Review*, passed this tradition, but measuring under the CIC system, to Richard Prior around 1959. Richard Prior maintained it unbroken up until 1995 and I took it on between 1995 and 2007. G. K. Whitehead was, for many years, Chairman of the International Trophy Commission, and he and Richard Prior together started another tradition, of measuring at the annual exhibition of the St Hubert Club of Great Britain.

When Richard Prior joined the Game Conservancy in the 1970s, he introduced trophy measuring at the CLA Game Fair to add interest to their stand. He struck the first medals in 1970 whilst with the Forestry Commission, and the first Game Conservancy Medals in 1977. When he became Technical Adviser to the British Deer Society, it was agreed that their name should be added to the measurement forms. Later on, the Game Conservancy withdrew, and the medals were re-struck in the BDS's name.

As demand for the service increased, Richard enlisted Allan Allison for Scotland in 1985 and myself for England in 1988. In 1996 Richard Prior passed his trophy measuring business to me 'lock, stock and barrel' but by 1997, as interest in the Grand European Exhibitions was beginning to wane, the CIC decided to form National Trophy Commissions and approached its UK Delegation to form a UK representation. My

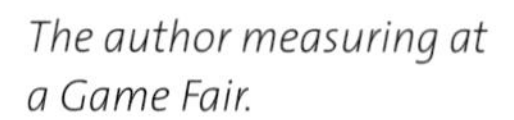

The author measuring at a Game Fair.

involvement as coordinator of the UK Commission continued until 2007. At the time of writing, I head up the BASC trophy measuring service, which is benchmarked to CIC measurements.

It is indicative of the success and the spread, both of the roe and muntjac, and of the two complementary measuring services, that the number of trophies measured continues to rise year on year. However, over the years since the first edition of this book, there have been interesting developments regarding the distribution pattern of quality roe. Whilst Hampshire, Wiltshire, Dorset, Berkshire and Somerset continue to dominate the lists of great heads, Somerset has leapfrogged two places over Berkshire and Dorset to claim third place in the list of gold medals. Most of the other counties have shown proportionate rises to their 2003 figures – although some of the very low numbers show statistically insignificant (though apparently high) rises.

Table 12.1 demonstrates how the statistics for gold medals have changed since 2003, with the figures up to that date compared to the figures up to 2010.

The table clearly shows that the five core counties (Hampshire, Wiltshire, Somerset, Dorset and Berkshire) remain the same and continue to produce excellent quality roe. In fact three-quarters of all recorded gold medals emanate from those five counties, with Hampshire alone producing around a quarter. However, quality roe are also clearly spreading, with Gloucester, Oxfordshire and Warwickshire demonstrating how that spread is progressing. Yorkshire and Northumberland are also now producing significant numbers of gold medals. In all, 22 English counties have contributed to the list, and 2005 appears to stand out as the best year on record.

Although some people disapprove of the very concept of measuring trophies, a significant sample of the premium trophies shot each year in the UK are measured, and one cannot therefore deny the enormous amount of data which this provides. Indeed, were it not for this system of recording, there would be absolutely no data on quality in roe over the past 40 years. It is true from a statistical point of view that it would be better if *all* trophies were recorded, but few would support the huge and unnecessary state control of stalking which would be required to administer this. I am satisfied that our voluntary system, now extended through the BASC system, offers us a large enough representative sample to reflect the bigger picture across the country. This means that any conclusions drawn from the data are both unique and important, and the key conclusion is that the management of roe must be extremely good across the country if the five core counties – which have, after all, been fully populated for around 20 years – continue to produce such fantastic world-beating quality.

In updating the list of the 100 strongest roe, it was also interesting to see that the benchmark for inclusion in the list has risen from 154.88 points to 158.16 points. Furthermore, there has been another massive development; the list shows that 54 per cent of the trophies emanate from

Table 12.1 Comparative list of gold medal trophies: 1991–2003 and 1991–2010

County	1991–2003	1991–2010	Change since 2003	
Hampshire	125	208	+83	166%
Wiltshire	93	173	+80	186%
Somerset	55	116	+61	211%
Dorset	59	90	+31	153%
Berkshire	41	72	+31	176%
Gloucester	27	51	+24	189%
Surrey	28	38	+10	136%
Sussex	20	37	+17	185%
Oxon	11	27	+17	245%
Yorkshire	10	16	+6	160%
Kent	5	6	+1	120%
Devon	6	11	+5	183%
Avon	2	3	+1	150%
Northumberland	2	9	+7	450%
Suffolk	2	3	+1	150%
Shropshire	2	2	—	—
Durham	1	1	—	—
Warwickshire	1	7	+6	700%
Cornwall	1	1	—	—
Cumbria	1	2	+1	200%
Essex	0	1	+1	100%
Herefordshire	0	2	+2	200%
Total	**491**	**876**	**+385**	

Table 12.2 Gold medal heads by county 1991–2009

County	91	92	93	94	95	96	97	98	99	2000	01	02	03	04	05	06	07	08	09	Total
Hampshire	12	13	9	10	6	8	15	3	9	5	7	13	15	16	19	20	9	11	8	208
Wiltshire	9	4	6	11	6	4	8	10	2	7	4	10	12	17	25	14	8	2	14	173
Somerset	2	6	1	1	0	3	3	1	2	8	3	10	15	10	12	14	6	8	11	116
Dorset	3	6	4	3	6	10	6	6	2	2	3	3	5	7	4	8	6	4	2	90
Berkshire	7	0	4	2	0	4	2	4	7	1	2	6	2	14	5	4	5	2	1	72
Gloucester	1	1	0	2	0	3	3	1	2	5	1	6	2	5	2	0	7	3	7	51
Surrey	4	1	9	2	1	2	1	4	1	1	1	1	0	0	3	2	3	1	1	38
Sussex	4	1	1	2	1	1	0	3	0	5	1	0	1	4	4	4	2	2	1	37
Oxon	1	1	1	1	0	1	1	1	1	0	2	1	0	7	1	3	2	2	1	27
Yorkshire	0	0	0	2	0	2	2	1	1	1	0	0	1	0	1	0	2	2	1	16
Devon	1	0	1	0	2	0	0	0	0	0	0	1	1	1	1	1	1	0	1	11
Northumbld	0	0	0	0	0	0	0	0	0	1	1	0	0	0	1	1	1	1	3	9
Kent	0	0	1	1	0	0	0	0	2	0	0	1	0	0	0	1	0	0	0	6
Warwick	0	0	0	0	0	0	0	0	0	0	0	0	1	0	1	2	0	2	1	7
Avon	0	0	0	0	0	0	1	0	0	0	1	0	0	0	0	0	0	0	1	3
Suffolk	0	0	0	0	0	0	1	1	0	0	0	0	0	0	0	0	0	0	1	3
Shropshire	0	0	0	0	0	1	1	0	0	0	0	0	0	0	0	0	0	0	0	2
Durham	0	0	0	1	0	0	0	0	0	0	0	0	0	0	0	0	0	0	0	1
Cumbria	0	0	0	0	0	1	0	0	0	0	0	0	0	0	0	0	0	0	1	2
Cornwall	0	0	0	0	0	0	1	0	0	0	0	0	0	0	0	0	0	0	0	1
Hereford	0	0	0	0	0	0	0	0	0	0	0	0	0	0	0	0	0	1	1	2
Essex	0	0	0	0	0	0	0	0	0	0	0	0	0	1	0	0	0	0	0	1
Total	**44**	**33**	**37**	**38**	**22**	**40**	**45**	**35**	**29**	**36**	**26**	**52**	**55**	**82**	**79**	**74**	**52**	**41**	**56**	**876**

the 2000s, 21 per cent from the 1990s, 16 per cent from the 1980s, 4 per cent from the 1970s and 3 per cent from the 1960s. Although frequency of measurement has clearly risen since the 1960s, it has not risen sufficiently in the last 20 years to explain away this significant and recent increase of huge heads. It simply must be the case that this is yet another demonstration of the excellence in deer management that is being pursued in much of England. I should add an apology for any errors or omissions in this list.

The UK record has been broken by chef Marco Pierre White (whose trophy also held the record for length at an average of 31.9cm) and there are five new inclusions in the top ten. In the non-typical list, there are now 17 trophies, two of which have exceeded Major Baillie's original 'Baillie Monster' of 238.55 points. Of those two, Tom Troubridge's was formally submitted, but rejected, as a potential new world record.

Mirroring the county profile of gold medals, perhaps unsurprisingly Hampshire produces almost a quarter of the top 100 trophies, with Wiltshire producing 18 and Somerset 13. Thereafter Dorset, Sussex, Berkshire, Oxfordshire, Surrey, Gloucestershire and Warwickshire all contribute a number of heads each, while Devon, Avon, Cumbria and Yorkshire each contribute one.

Marco Pierre White's UK record buck.

It is also interesting to note that there is welcome consistency of measurement, which reflects the importance of weight in achieving a high score. Dividing the list into equal fifths of 20 trophies each, the average weights reduce in a predictable and reasonable progression:

Top 5th	716g
Second 5th	649g
Third 5th	618g
Fourth 5th	606g
Bottom 5th	594g

The list of strongest muntjac trophies shows a similar result. Back in 2003, a total of only 82 trophies over 65 points were listed, but now I am able to list 100 trophies and the benchmark for inclusion has risen to 66.6 points. The UK record has been broken by J. Harding's 2009 Buckinghamshire buck of 79.9 points and, as with Marco Pierre White's record roe, this muntjac trophy also sets a record for the longest beams, at an average of 17.45cm. The longest brow points, at an average of 4.55cm, are attributable to P. Lever's Berkshire buck of 2007, and the widest head at 15.1cm remains that of L. Price's Buckinghamshire buck of 1997.

Few people would have a problem with the concept of measuring and recording as a guide to assessing trends in quality. Our problem is perhaps in the issue of medals. Some see this as encouraging an unhealthy competitiveness, or even as in some way rewarding greedy stalkers. This underestimates the enormous enthusiasm for the species which most roe stalkers expound. Whether this enthusiasm would extend to voluntarily submitting trophies for assessment and recording, were it not for the lure of that little medal, it is difficult to know. In any event, I believe that most stalkers regard the medal not as a personal achievement, but as a tribute to a well-loved old buck.

Stalkers were certainly shooting medal trophies before 1960, when formal recording began. Others were being taken in winter drives, when their antlers were either cast or in velvet, and can therefore never appear in the records. However, few of these early trophies appear to be exhibited today, and most of the old-timers accept as fact that far more have been shot since the 1970s, than had ever been shot before. Therefore it is not just because of an increased interest in trophy measurement that the number of trophies measured per year has increased at a greater rate than

LEFT *M.J. Langmead, 1971 Sussex, a UK roe record for 35 years.*

ABOVE *K. Hicks, 79.5 points; No. 2 in the UK.*

BELOW LEFT *M. Giles, 77.3 points; No. 4 in the UK.*

BELOW *Blue Thomas with his UK No. 12 muntjac trophy.*

colonisation, and the number of medals presented today is showing absolutely no sign of decline. Furthermore, because the majority of silver and gold medals are from older bucks (estimated at over five years of age), there is a valuable incentive for stalkers to leave the good young middle-aged bucks, in the hope that they might develop further. This ensures a well-structured population in line with best theoretical management practices.

Brett, 200.5.

D. Hunt, 168.65.

J. Bundy, (non-typical) 228.78

V. Maguire, 191.95.

In conclusion, the current position is very healthy, and challenges continuing assertions that there are too many deer which are not being properly managed. In the main, our roe are well managed by a dedicated and enthusiastic body of recreational and professional stalkers alike.

P. Reed, 173.6.

N. Dewing, 170.2.

Mr Whatley, 151. 85.

J. de Lobel, 187.85.

M. Edwards, 188.6.

A. Robinson, 175.13.

Non-typical, S. Harcourt, 218.15.

J. Crockford, 157.8.

G. Tunal, Sussex, 160. Photo by G. Tunal

M Sworder, 157.

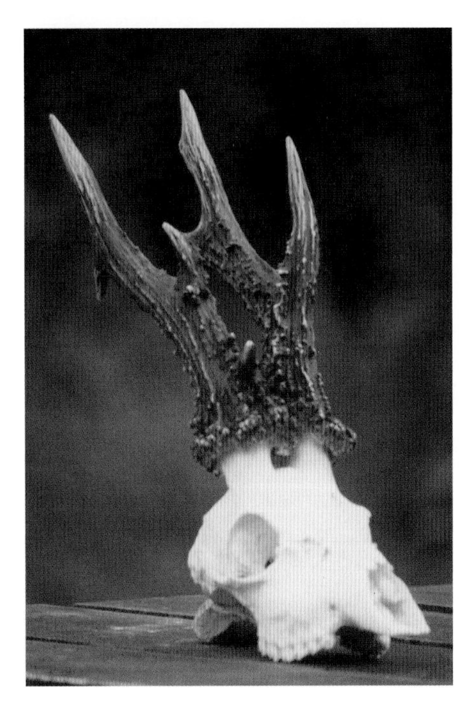

A. Shannahan, 169.65.

D. Graham 192.2

Table 12.3 England's strongest 100 roe trophies 1960–2010

Name	County	Date	Points	Av. Lgth (cm)	Weight (g)	Score
M.P. White	Hants	04.06	6	31.90	930	222.65
M.J. Langmead	Sussex	04.71	12	26.30	750	210.25
D. White	Warwk	07.06	7	25.35	774	205.18
K. Brett	Glos	04.96	6	30.60	755	200.50
E. Moorhouse	Hants	05.83	6	27.35	—	197.18
O. Lindemann	Wilts	07.81	12	27.75	—	192.50
D. Graham	Hants	08.04	6	26.40	705	192.20
V. Maguire	Hants	05.02	9	27.90	700	191.95
A. Ayris	Wilts	08.06	7	26.00	765	191.50
B. Madsen	Somerset	09.08	8	28.35	750	190.98
M. Edwards	Hants	07.02	8	26.60	752	188.60
J. de Lobel	Berks	05.98	6	27.70	650	187.85
A. Ayris	Wilts	08.05	9	25.70	640	185.85
C. Jackson	Wilts	09.06	7	27.05	762	185.53
A. Thomsett	Sussex	2008	—	28.95	710	181.43
K. Rolfe	Berks	04.86	7	25.25	702	180.53
F. Lange	Sussex	1992	6	26.45	689	179.23
T. Malkin	Surrey	05.97	6	27.85	587	178.53
M. Morphett	Oxon	04.96	6	26.25	595	178.13
D.N. Carr-Smith	Sussex	08.69	6	26.00	589	177.80
A. Wilson	Wilts	10.04	6	27.60	715	177.30
P. Griffin	Wilts	06.91	6	27.15	682	176.88
H. Janssen	Hants	08.83	7	27.40	—	176.60
P. Conde	Somerset	2002	6	26.30	654	175.95
A. Roberts	Wilts	06.03	6	27.50	710	175.25
A. Robinson	Wilts	05.00	5	26.25	635	175.13
P. Bundy	Wilts	RTA	10	24.05	644	174.13
G. Jacobs	Dorset	05.81	6	26.55	—	173.80

table continues overleaf

Name	County	Date	Points	Av. Lgth (cm)	Weight (g)	Score
P. Reed	Hants	04.02	14	27.20	670	173.60
I. Hockley	Sussex	08.91	6	28.75	590	172.88
G. Dubaere	Oxon	07.04	6	24.70	715	172.85
R. Routledge	Oxon	2009	—	27.80	600	172.40
J. Lee	Dorset	06.96	13	24.40	680	171.20
M. Swan	Hants	08.87	6	28.30	610	171.15
J. Layzell	Sussex	05.83	6	27.30	—	171.15
N. Dewing	Berks	06.02	6	26.40	610	170.20
A. Shannahan	Hants	09.96	7	28.30	680	169.65
I. Ballard	Glos	05.03	6	28.70	645	169.35
I. Parkin	Berks	06.04	8	26.60	575	168.80
D. Hunt	Berks	08.02	13	23.30	620	168.65
I. Milne	Devon	04.93	8	24.35	610	167.68
R. Hann	Somerset	06.07	6	27.40	655	166.70
S. Caddy	Wilts	08.91	6	27.00	642	166.40
J. Jones	Somerset	2002	6	25.95	740	166.28
T. Hehir	Hants	2009	—	22.40	620	166.20
C. Von Langen	Dorset	05.82	10	28.55	—	166.10
J. Duddy	Surrey	06.93	6	31.45	588	166.03
R. Rashidian	Dorset	04.95	5	26.50	680	165.75
K. Franklin	Warwk	09.06	8	24.45	615	165.73
M. Van Der Moortelle	Surrey	08.06	7	28.80	594	165.70
K. Trinder	Surrey	1984	6	24.25	—	165.15
M. Broad	Hants	2002	7	28.65	605	164.83
R. Hughes	Berks	08.98	6	24.85	583	164.63
L. Ritchie	Somerset	08.96	6	26.00	600	164.50
C. King	Somerset	05.06	6	27.70	600	164.35
E. Moorhouse	Hants	05.89	7	26.05	588	164.23
C. Oliver	Avon	1985	8	25.45	555	164.23
M. Bregolli	Hants	06.79	6	27.20	—	164.20

Name	County	Date	Points	Av. Lgth (cm)	Weight (g)	Score
S. Sparrow	Dorset	RTA	6	24.65	630	163.83
T. Boffy	Hants	10.95	9	24.30	630	163.65
M. Bright	Hants	05.05	8	24.05	655	163.53
N. Parker	Hants	08.06	6	27.50	500	163.15
B. Dewitte	Wilts	05.03	6	29.00	590	163.00
J. Jones	Somerset	2002	6	25.55	655	162.78
R. Burleigh	Wilts	07.06	8	26.65	620	162.63
E. James	Dorset	05.78	6	29.15	—	162.60
D. Hunt	Hants	05.03	6	29.00	600	162.50
J. Maunder Taylor	Hants	10.04	6	27.95	603	162.18
B. Andrews	Wilts	04.07	7	24.05	632	162.03
R. Jones	Wilts	04.05	7	25.35	592	161.98
D. Cailes	Dorset	04.03	6	26.50	680	161.75
P. Prince	Dorset	05.00	6	24.45	600	161.63
W. Witchell	Glos	04.07	9	28.50	610	161.25
M. Bregolli	Hants	09.81	7	25.40	—	161.20
C. Rowland	Wilts	04.03	6	27.30	555	161.15
J. Hobbs	Oxon	04.03	9	26.30	565	161.15
K. Al-Tajir	Sussex	10.90	7	24.15	601	161.14
B. Jones	Oxon	2001	6	25.00	620	161.00
G. Tunal	Sussex	07.00	7	24.45	622	160.93
S. Fancy	Somerset	07.99	7	28.55	610	160.78
J. Emssens	Somerset	08.04	6	26.05	610	160.53
P. Fisher	Hants	06.05	6	26.05	610	160.53
E. Slater	Sussex	1964	4	26.10	530	160.50
J. Turner	Dorset	05.86	10	23.95	600	160.48
C. Wilp	Somerset	07.01	6	26.85	600	160.43
S. Philipps	Somerset	05.69	6	27.90	—	160.22
D. Cox	Somerset	04.05	8	26.90	630	159.95
J. Keyl	Hants	05.81	6	29.55	—	159.90

table continues overleaf

Name	County	Date	Points	Av. Lgth (cm)	Weight (g)	Score
L. Petyt	Wilts	05.77	8	27.76	—	159.85
P. Bond	Somerset	06.03	6	23.70	680	159.85
R. Haider	Hants	05.00	6	26.15	580	159.63
E. Brown	Dorset	05.06	6	23.90	668	159.55
D. Horton	Wilts	06.80	6	26.00	—	159.50
T. Lawrence	Cumbria	06.83	6	28.90	—	159.45
S. Jones	Hants	07.06	6	25.80	587	159.20
W. Morlins	Hants	07.93	6	28.85	535	158.93
A. Mead	Wilts	05.05	8	29.50	570	158.75
M. Jury	Dorset	07.93	6	27.80	603	158.60
G. Harris	Sussex	07.99	7	27.15	562	158.38
R. Kramer	N.Yorks	08.08	6	27.45	545	158.16

Table 12.4 Top non-typical trophies

P. Howard	Hants	Fnd Dead	7	23.6	1055	289.3	Bailliesque
T. Troubridge	Dorset	05.06	8	29.9	1182	275.65	Bailliesque
Maj.Hon.P Baillie	Hants	05.74	6	21.25	1032	238.55	Orig. Baillie Monst.
J. Pilkington	Wilts	05.91	8	22.5	990	236.25	Bailliesque
J. Bundy	Wilts	c1970	6	28.75	955	228.78	Bailliesque
Cowdray Estate	Sussex	Fnd Dead	9	24.9	695	228.45	Perruque
P. Dalton	Wilts	Fnd Dead	7	24.85	975	222.93	Bailliesque
S. Harcourt	Oxon	2001	6	23.3	990	218.15	Bailliesque
D. Barton	Hants	Fnd Dead	10	23.7	805	216.85	Bailliesque
M.P. White	Hants	1994	4	27.05	862	215.33	Bailliesque
S. Whitfield	Berks	04.87	4	27.15	896	214.88	Bailliesque
D. Andress	Hants	08.92	5	27.7	742	208.55	Bailliesque
V. Pardy	Hants	Fnd Dead	6	20.6		205.30	Bailliesque
BDS	Cornwall	Fnd Dead		20.1	700	192.05	Perruque
M.P. White	Somerset	07.10	6	21.6	705	188.30	Perruque
P. Vastapane	Hants	04.99	6	23.3	780	174.15	Bailliesque
G. Lewis	Wilts	1991	6	22.35	680	161.48	Bailliesque

County totals top 100

Hants	24	Oxon	5	Yorkshire	1
Wilts	18	Surrey	4	Cumbria	1
Somerset	13	Glos	3		
Dorset	11	Warwickshire	2		
Sussex	10	Avon	1		
Berks	6	Devon	1		

'60s 3 '70s 4 '80s 16 '90s 21 '00s 54 Unknown 2

Photo by Brian Phipps

Table 12.5 England's top 100 muntjac

						Beams		Brow		
Name	**County**	**Date**	**Age***	**Pts**	**Span**	**L**	**R**	**L**	**R**	**Score**
J. Harding	Bucks	Feb 09	old	4	12.7	17.4	17.5	2.3	1.9	79.9
K. Hicks	Glos	Oct 97	ma	4	12.5	14.0	14.0	3.8	3.3	79.5
R. Birt	Hants	Aug 03	old	4	1.3	12.7	13.0	2.6	3.1	78.3
M. Giles	Hants	Feb 98	old	4	11.5	15.3	15.0	3.4	3.3	77.3
P. Childerley	Cambs	Oct 09	lma	4	12.8	14.0	14.9	1.6	1.6	76.6
S. Whitelock	Beds	Feb 98	ma	4	13.1	13.6	14.7	2.8	2.3	75.1
L. Crook	Beds	2009	old	4	13.1	13.9	14.8	3.0	2.0	75.0
D. Hunt	Wilts	2009	old	4	11.1	15.1	14.8	2.1	1.9	74.5
O. Salven	Oxon	Dec 01	old	4	12.6	14.7	13.5	2.3	1.6	73.8
Bedford Estate	Beds	Jan 01	old	4	12.3	15.1	15.9	2.6	2.4	73.7
T. Magee	Berks	May 06	old	4	10.4	16.5	16.0	2.1	1.9	73.6
M. Thomas	Oxon	April 98	ma	4	13.8	13.4	13.2	1.3	1.7	73.4
S. Freedman	Oxon	May 98	old	4	10.6	14.5	14.4	2.5	2.1	73.4
B. Sherriff	Herts	Aug 99	lma	4	11.7	15.2	16.0	1.7	1.3	73.3
A. Giles	Berks	April 07	—	4	11.8	14.4	15.1	2.7	2.2	73.2
C. Smith-Jones	Herts	Feb 04	lma	4	10.6	13.3	12.9	3.2	3.3	72.8
I. Miller	Glos	Oct 96	—	6	13.6	12.5	12.4	3.6	3.9	72.8
C. Agg	Oxon	Mar 95	ma	4	12.1	13.8	14.6	2.8	2.5	72.7
R. Wakefield	Oxon	Nov 99	old	4	12.3	13.4	13.1	2.2	1.9	72.5
P. Easeman	Bucks	Aug 07	old	2	12.2	16.1	14.9	1.8	1.6	72.4
G. Frost	Berks	Sept 01	old	4	12.4	15.5	15.1	1.2	1.4	72.2
R. Everett	Oxon	1995	ma	4	10.7	14.1	14.1	3.3	2.7	72.1
J. Guyas	Middx	Nov 97	old	4	12.4	14.6	14.2	1.7	2.3	72.0
K. Siford	Hants	Nov 08	old	4	12.7	13.9	13.4	2.9	3.0	71.7
M. Mercer	Beds	Feb 98	ma	5	11.6	14.4	13.8	1.9	1.5	71.6

						Beams		Brow		
Name	County	Date	Age*	Pts	Span	L	R	L	R	Score
M. Seare	Herts	Feb 96	old	4	12.6	13.3	13.2	1.9	2.5	71.6
R. Simmons	Berks	May 06	lma	4	11.4	15.4	16.0	2.7	2.3	71.6
P. Easeman	Bucks	Mar 08	old	4	12.0	15.4	14.6	1.7	1.6	71.6
A. de Ryck	Berks	2008	old	4	10.4	14.5	13.9	2.3	2.2	71.4
W. Wycombe	Bucks	Aug 01	ma	4	10.4	14.8	14.4	2.9	2.3	71.2
M. Stearns	Berks	Feb 99	lma	4	13.5	14.7	14.3	2.1	2.7	70.9
A. Hogg	Hants	Aug 99	old	4	11.7	14.0	13.6	2.0	2.1	70.7
K. Dixon-Nutt	Bucks	Sept 93	old	4	13.5	13.0	12.7	1.9	1.4	70.7
M. Westall	Berks	Mar 02	ma	4	12.8	14.6	15.0	1.4	1.2	70.6
L. Beerden	Berks	2008	lma	4	12.8	13.4	12.4	2.5	2.4	70.6
J. Meaker	Hants	May 02	ma	4	11.5	13.1	13.5	2.6	2.7	70.5
D. Stacey	Somerset	1999	ma	4	9.7	11.7	12.5	4.0	3.0	70.5
A. Robinson	Berks	Sept 01	old	4	13.4	13.2	13.5	2.7	2.4	70.4
T. Parr	Norfolk	Nov 95	ma	4	14.4	13.5	14.0	1.9	1.4	70.2
Lt Col. Lorimer	Hants	1986	lma	4	12.3	12.8	13.0	—	—	70.1
M. Duggan	Herts	1996	lma	2	13.5	16.4	15.6	—	—	70.1
S. Little	Hants	Oct 02	old	4	10.9	12.5	12.2	1.5	1.5	70.1
B. Johnson	Oxon	Jan 97	ma	4	11.4	14.0	14.9	2.0	2.4	70.0
J. Childs	Berks	1995	old	5	11.6	12.5	14.0	2.3	2.5	70.0
N. Bunce	—	1994	ma	2	11.4	13.4	13.7	—	—	70.0
G. Walker	Oxon	Feb 01	old	4	12.9	14.6	16.0	1.2	1.5	69.9
P. Lever	Berks	Sept 07	—	4	10.6	15.5	15.0	4.6	4.5	69.8
R. Knight	Beds	Mar 00	old	5	14.4	12.8	13.7	1.5	1.3	69.6
R. Simmons	Berks	Aug 03	ma	4	11.8	14.0	14.4	2.4	1.7	69.6
K. Cook	Feb 09	Suffolk	old	4	12.5	13.6	12.9	2.0	2.2	69.5

table continues overleaf

						Beams		Brow		
Name	County	Date	Age*	Pts	Span	L	R	L	R	Score
R. Simmons	Berks	July 06	old	4	13.1	13.6	12.5	1.5	1.5	69.3
P. Goddard	Bucks	Feb 05	old	4	10.3	14.4	13.6	1.2	1.7	69.3
L. Price	Bucks	Aug 97	ma	2	15.1	13.4	14.8	—	—	69.3
D. Pike	Wilts	Jan 09	old	4	13.8	12.6	12.8	1.5	1.7	69.2
P. Van Jensen	Berks	April 99	old	4	12.9	13.9	13.9	1.3	1.4	69.1
S. Hume	Oxon	2007	—	4	11.7	13.2	11.9	1.3	1.4	69.1
T. Laws	Lincs	Feb 07	—	5	10.7	12.8	13.0	2.2	2.6	69.1
C. Perkins	Norfolk	May 06	old	6	11.0	10.6	11.2	3.0	3.3	69.0
R. Birt	Hants	2002	lma	4	13.9	13.8	14.0	1.9	2.2	68.9
A. Alberici	Herts	Aug 06	old	4	12.7	12.9	13.3	1.9	1.5	68.7
C. Smith-Jones	Berks	Oct 06	—	6	11.3	13.2	12.3	3.1	3.4	68.7
M. Sayfritz	Berks	Mar 06	—	4	10.4	12.0	12.0	1.7	2.1	68.6
P. Easeman	Bucks	Jan 01	old	4	11.8	13.6	14.4	1.7	1.6	68.6
J. Curtis	Oxon	Jan 02	ma	4	10.6	13.9	13.2	2.4	2.0	68.6
M. Giles	Hants	Oct 97	old	4	10.7	13.8	14.3	1.9	1.9	68.5
R. Ward	Berks	Aug 01	ma	4	10.8	13.9	13.5	3.2	3.4	68.4
T. Hughes	Berks	Aug 94	old	4	10.8	13.5	13.6	3.5	2.8	68.4
J. Harding	Bucks	2006	—	4	13.0	12.2	12.2	3.1	2.7	68.4
P. Hurn	Norfolk	—	old	4	12.6	13.0	13.1	1.5	1.0	68.4
A. Fitch	Northants	Oct 09	old	4	10.7	14.5	14.1	1.4	1.0	68.3
P. Hughes	Northants	1993	ma	4	12.1	12.0	12.4	2.0	1.8	68.1
R. Spence	Huntingdn	1999	ma	4	12.6	13.2	12.7	2.2	2.1	68.0
Dr Newton	Oxon	1980	ma	4	10.2	13.5	13.9	—	—	68.0
M. Webber	Beds	Jan 06	old	4	10.0	12.1	11.9	1.9	1.9	68.0
J. Wasylowski	Oxon	—	—	4	13.3	11.0	11.3	1.9	1.9	67.9
A. Chamberlain	Staffs	2010	ma	4	12.9	11.1	11.2	2.4	2.4	67.9
A. Lord	Beds	Apr 07	—	4	11.4	12.0	12.0	4.5	4.3	67.8

						Beams		Brow		
Name	County	Date	Age*	Pts	Span	L	R	L	R	Score
H. Bulder	Warwick	Apr 03	ma	4	11.6	11.5	13.6	2.4	2.8	67.6
C. Halls	—	Dec 99	ma	4	10.0	13.0	14.5	2.6	2.5	67.6
P. Goodey	Herts	2000	old	4	11.7	12.4	10.7	2.2	2.4	67.5
A. Asher	Beds	2008	old	4	12.0	14.0	14.7	1.7	1.1	67.5
A. Curd	Herts	Feb 04	ma	4	11.9	14.6	14.7	2.3	1.3	67.5
J. Flynn	Worcs	26.04.00	old	4	12.4	12.1	12.3	2.2	2.6	67.4
R. Hermann	Oxon	Feb 95	old	4	11.8	13.3	13.7	1.5	1.2	67.4
P. Howard	Oxon	Mar 06	old	4	13.6	13.9	13.4	1.3	1.8	67.2
Mr Hedges	Northants	1987	ma	4	10.7	13.4	13.0	—	—	67.1
E. Penser	Wilts	1995	ma	4	11.7	12.8	12.5	2.8	2.9	67.1
A Meakin	Northants	Aug 96	ma	4	10.8	12.4	11.7	2.3	2.0	67.1
J. Dow	Beds	Feb 06	old	4	13.1	12.4	10.2	2.6	2.6	67.0
A. Patmore	Northants	Feb 97	old	4	11.0	13.8	14.4	1.8	1.4	67.0
B. Scott	Cambs	Mar 06	—	4	11.4	12.7	14.0	2.3	2.3	67.0
I. Parkin	Bucks	Feb 99	old	4	13.2	12.3	13.2	1.7	1.7	67.0
J. Savage	Cambs	Dec 99	ma	4	13.3	13.5	12.8	1.4	1.4	66.9
D. Walker	Hants	1996	lma	4	11.6	13.5	13.5	1.8	1.2	66.8
R. Simmons	Berks	Sept 05	ma	4	11.4	12.9	11.7	2.0	2.1	66.8
O. Beardsmore	Oxon	May 05	old	6	12.6	13.0	12.5	2.5	1.7	66.8
N. Harwood	Northants	Dec 04	old	4	12.5	11.4	10.8	2.3	2.5	66.7
C. King	Northants	Sept 96	lma	4	13.1	12.9	12.2	2.0	5.5	66.7
C. Allen	Berks	Sept 09	old	4	11.3	11.6	11.5	2.8	2.2	66.7
K. Mckenchie	Bucks	Oct 08	old	4	11.0	13.5	13.4	1.4	1.5	66.6

* ma = middle aged; lma + late middle aged

13.

MY EIGHT GREATEST ENGLISH ROE TROPHIES

IN THE 50-ODD years since we have been recording roe trophies, there have been many significant milestones. There have been very large heads, bizarre heads, non-typical heads and controversial heads. These are my top eight, a personal selection, in the order of appearance. Most are recent and I make no apology for this. That said, for me it started in 1971.

1. The Langmead head

Name	Date	County	No. tines	Av. lgth (cm)	Net weight (g)	Score
M. J. Langmead	April 1971	Sussex	12	26.3	750	210.25

In 1971, Michael Langmead's trophy was in a class quite of its own. Enormous, unusual, unprecedented, for 35 years it remained the UK record roe buck. It was shot in Sussex when this was the pre-eminent county for roe, but one which has now sadly declined. It was measured at the Plovdiv Trophy Exhibition (the last international exhibition) of 1981 and sports that medal, together with those of the Game Conservancy and St Hubert Club. I visited Mr Langmead in 2003 just before first publishing this book. This trophy remains the one which, of all, I would most like to exhibit on my own wall. It has everything that a roe trophy should have.

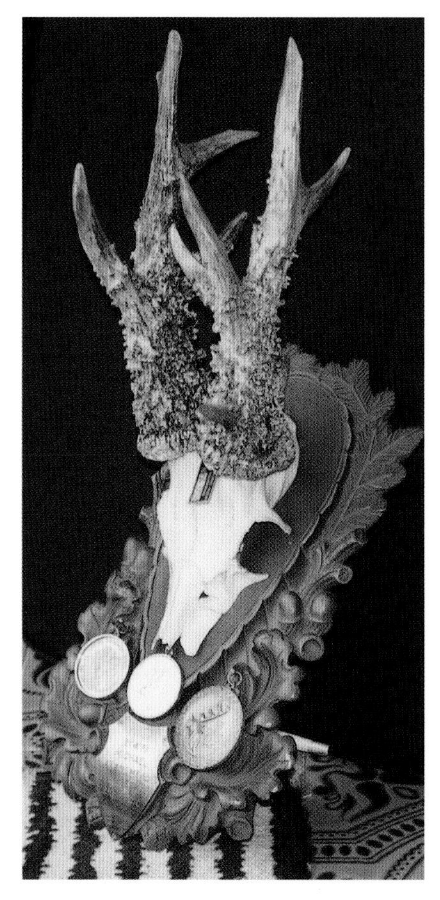

The Langmead head.

2. The original 'Baillie Monster'

Name	Date	County	No. tines	Av. lgth (cm)	Net weight (g)	Score
Maj. Hon. P. Baillie	May 1974	Hants	6	21.25	1032	238.55

Major Baillie's trophy of 1974 must be, in many respects, the most important head in the history of trophy assessment as it represents a benchmark in what became accepted or rejected for official measurement. Measured as a world record, but subsequently disqualified, it gained the now well-known epithet of 'Baillie Monster'. Although huge of antler, it exhibited unusual and excessive bone growth with 'the appearance of poured and hardened cement' around the coronet and extending to the underside of the eye socket. International judges decided that this made it non-typical and unfair to judge against 'normal' trophies. At the time it was unique, but over the years more have been presented and there are now at least 20 of its type known to exist amongst collections. Why they develop remains a mystery, but they are associated with areas where very high-protein food is in abundance (low-ground pheasant shoots). More than half a dozen have passed through my hands over the years, and my only comment is that if a 'Baillie Monster' were to be judged a world record it is a fact that no 'normal' trophy could ever retake that title.

3. The current world record

Date	County	No. tines	Av. lgth (cm)	Net weight (g)	Score
1982	Sweden	8/9	26.8	875	246.90

This is the only non-UK trophy which appears in this list and it is here because it represents a benchmark against which all others must be judged. Completely normal in skull, it sports enormous and, to my mind, beautiful and unusual antlers. Recently challenged by the Troubridge trophy (see below) it remains the world record. I am extremely lucky to own an original copy of this head and it remains the pride of my personal collection.

4. The Brett head

Name	Date	County	No. tines	Av. lgth (cm)	Net weight (g)	Score
K. Brett	April 1996	Gloucester	6	30.6	755	200.50

Ken Brett is a South African who was visiting the UK. Out stalking with a colleague in Gloucestershire, he was extraordinarily lucky to have the opportunity to shoot this buck. It was the first trophy since 1971 which challenged Michael Langmead's UK record and created considerable excitement at the 1996 Game Fair. At final analysis, however, it fell a full 10 points short and took second place behind that trophy. Today it retains fourth place.

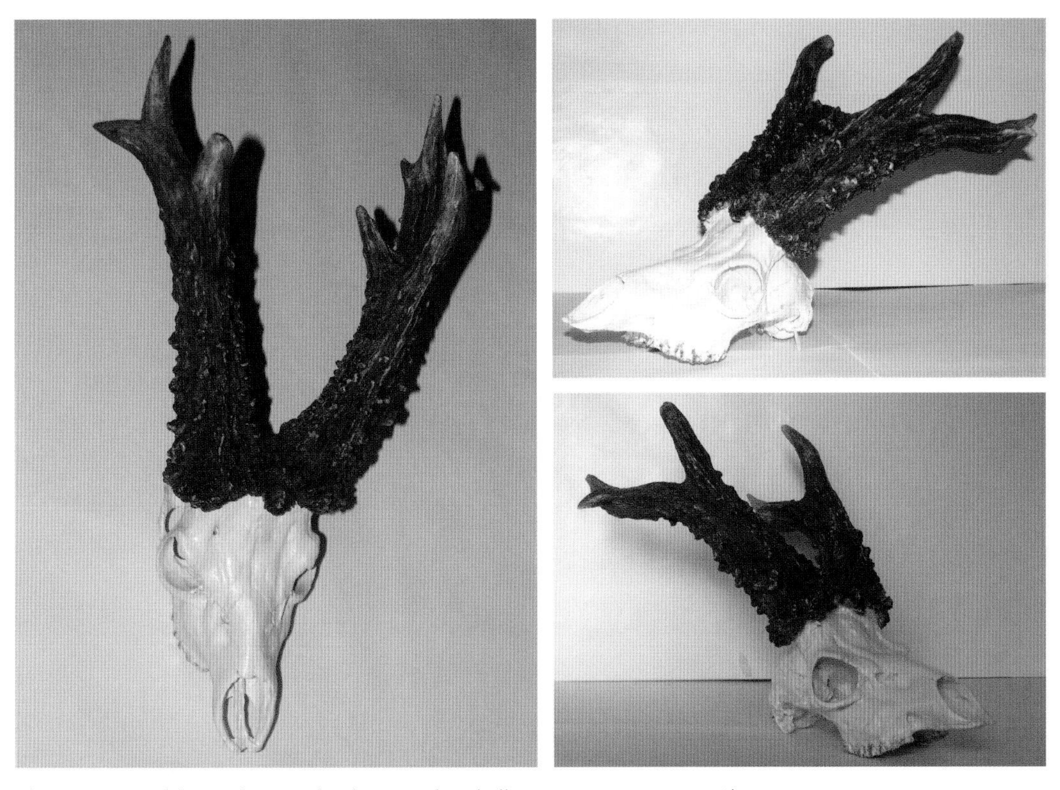

The current world record – completely normal in skull, it sports enormous antlers.

Ken Brett's was the first trophy since 1971 to challenge Michael Langmead's UK record.

5. The current UK record

Name	Date	County	No. tines	Av. lgth (cm)	Net weight (g)	Score
M Pierre White	April 2006	Hants	6	31.9	930	222.65

Marco Pierre White is a passionate roe enthusiast who probably spends as much time stalking as many professional stalkers. In March 2006, while he was looking at the bucks at Broadlands with Bill Webb, an enormous buck, half out of velvet, crossed the road in front of them. As it jumped up the bank and was silhouetted against the setting sun, it was clear that they had found a monster. Some weeks later a farm contractor saw it and phoned to report that it could be found sharing a grass field with the cows. It was actually shot inside the adjoining wood after a difficult stalk in thick undergrowth. It was officially measured at the CLA Game Fair (which by chance happened to be at Broadlands that year) and declared the new UK record. By coincidence it was also the longest recorded head.

Marco Pierre White's current UK record.

6. The Troubridge head

Name	Date	County	No. tines	Av. lgth (cm)	Net weight (g)	Score
T. Troubridge	May 2006	Dorset	8	29.9	1182	275.65

Quite extraordinarily, after a 35-year reign, two bucks came along at the same time to challenge Michael Langmead's UK record. Mr Tom Troubridge was lucky enough to shoot this absolutely massive buck. Exceeding that of Marco Pierre White by some 50 points it was a contender for the title of new UK and world record. However, despite being initially accepted by members of the UK Commission, it was disqualified by the International Trophy Commission which stated: 'The International Trophy Measuring Commission has come to the conclusion that the

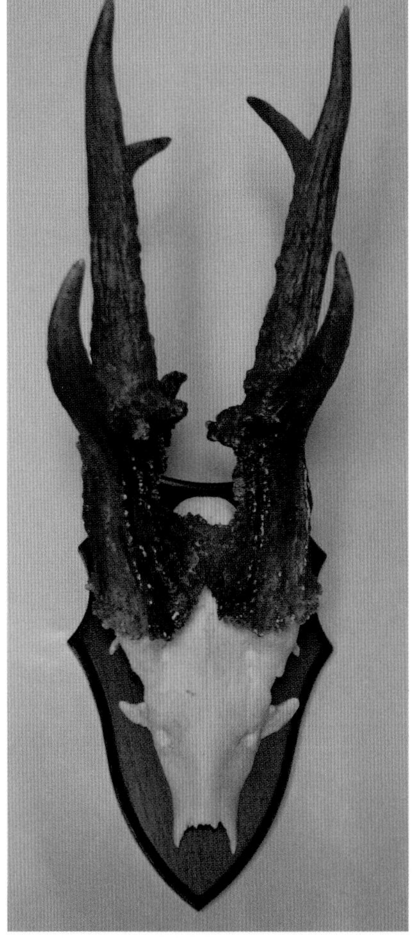

Tom Troubridge's massive buck.

measuring of this roe buck trophy is not objective, as the weight is mainly contained in the skull mass. The formation of the skull is abnormal.' It therefore remains classified as non-typical. It is nevertheless the second-largest trophy ever measured in England, and almost certainly the largest ever shot. To the purist, however, it will be something of a relief that this trophy was rejected because one thing is certain: had this trophy been declared a world record, no 'normal' trophy could ever have regained that title.

7. The road traffic accident

Name	Date	County	No. tines	Av. lgth (cm)	Net weight (g)	Score
P. Howard Found dead	2006	Hants	7	23.6	1055	289.30

This buck was found dead in a field and was quite decayed. The mice had started to gnaw the antler and it was in a sorry state. Scoring 289.3 it is the largest-ever recorded roe and one can only imagine how it would have scored as a clean-boiled fresh trophy. As another 'Baillie Monster' it

Found dead, this buck was the largest-ever recorded roe.

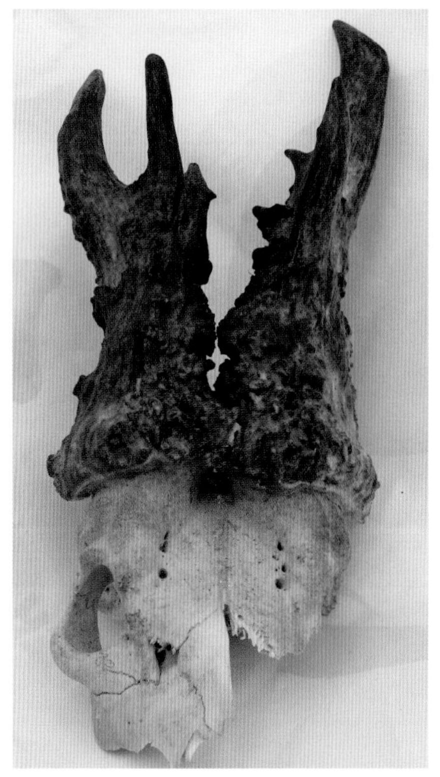

is, of course, ineligible for official recognition, but as a trophy it remains a highly sought-after collector's item. Leading taxidermist, Colin Dunton has done a magnificent job of setting it up and it takes pride of place within its owner's collection.

8. The double trophy

Name	Date	County	No. tines	Av. lgth (cm)	Net weight (g)	Score
M. Pierre White	July 2010	Somerset	6	21.6	705	188.3

For many stalkers, the perruque is a lifelong ambition, being extremely rare and usually short-lived. Having taken the UK record in 2006 it is extraordinary that the opportunity to shoot a perruque should fall to Marco Pierre White. When he shot this perruque buck, it was only days from an agonising death. Fly-strike had started and the maggots would soon have tormented it beyond bearing. Colin Dunton has developed a

BELOW *Marco Pierre White's perruque: and above, details of the perruque under its velvet.*

unique technique to create a facsimile copy of the original using its original velvet and skin to make a shoulder mount, whilst leaving the skull and antlers as a secondary trophy. This method of preparation has proved at last that a so-called 'mossed head' is no more than a perruque which has died and from which all the living material has rotted. I had always suspected this, but the photos clarify the situation once and for all. Scoring 188.3, it represents Marco Pierre White's seventh gold medal in a single season.

14.

THE PRINCIPLES OF TROPHY MEASUREMENT AND MEASURING

THE MEASUREMENT of trophies provides invaluable information about the quality of deer on these Islands. Measurements are designed to favour the antlers of fully mature or older animals and there is therefore no avoiding the fact that, if we were overshooting or mismanaging our deer, average antler quality would diminish and we would recognise that through the annual returns.

The measurement of all the deer species involves physical measurements being taken of various aspects of each trophy. These physical measurements may then be scaled up or down using a multiplier according to what is considered to be their relative importance to an outstanding trophy of its type. For example, we would all accept that it is fairly common for a roe buck trophy to exhibit length, so just 10 per cent of the total score is attributed to length. What is clearly outstanding for a roe is to exhibit weight and volume, so these measurements are loaded to the extent that they account for about 75 per cent of the total score. Evenness, span, tine length and girth are considered important in muntjac and sika, while length, girth, weight and the number of points are important to red deer. Palmation is crucial to the fallow, of which the length and width alone account for up to 50 per cent of the overall score. In Chinese water deer, only the length and circumference of canine teeth are taken into account. Yes, there are some silly anomalies which should really have

been ironed out years ago – why should a roe trophy have an ascending and stepped reward for span up to 75 per cent of its average length, and then suddenly score zero for span in excess of 75 per cent? Surely a graduation both up and down would be more logical. Why are red tines measured differently from sika tines? Why should the muntjac, which has evolved as a species of dense, low forest, be rewarded for wide span when a narrow span is better suited to its natural environment? And yes, we have all got our views on the so-called beauty points – personally, I am very content with the roe formula and do not see that much margin, if the rules are applied as written, for interpretation between measurers. In fact, I really like the roe formula and very much hope that the planned changes don't happen. Perhaps the only glaring fault relates to the Chinese water deer in which only the canines are measured. Sadly, when the formula was developed, I don't think much was known about the natural history of this species. The tusks develop over 18 months and are at their peak at that stage of development. Thereafter, the effect of wear and damage through fighting causes the tusks to decrease in size. In this respect, therefore, it is the only formula which favours younger animals and, as such, does not fit comfortably with the ethic of trophy measurement.

Some trophies which do not require an evaluation of weight, such as sika and muntjac, may be provisionally measured fresh, clean-boiled or even fully mounted. In the case of the latter, assurances are sought that the taxidermist has used the original skull plate. Others (red, roe and fallow) require the skull to be weighed and can only therefore be measured clean-boiled and dry. A high standard of trophy preparation is desirable, and is, indeed, a requirement if accurate measurements are to be taken. Many roe shot in April and early May retain residual velvet, and although many stalkers prefer to leave their trophies in their 'original' state, all velvet should ideally be removed before a measurement is made as this could affect the volume measurement.

Measurements were designed to be made using a narrow, flat, flexible steel tape and suspension balance. Length measurements are usually taken to the nearest 0.1cm, and it is good practice to take each measurement three times to ensure accuracy. Circumference measurements should be taken by looping the tape around the antler, and passing the scale over the zero point before reading off the measurement. A tape with a run-off before the zero point will be required (see Equipment for trophy

Photo opposite by Brian Phipps

measuring). In the past, weight measurements were normally taken to the nearest 5g (roe) and 10g (red and fallow), but the existence of much more accurate scales appears to have led to a greater degree of accuracy. In any event, care should be taken to ensure that the balance is correctly calibrated before each measurement. 'Beauty points' are taken to the nearest 0.5 of a point. Despite my comments above, some variation in both physical and judgemental scoring is inevitable according, for example, to how the tape is handled, how the scale is handled or how the individual beauty points are assessed. Volume measurements in roe can be extremely challenging, particularly if the coronets are deeply sloping, in which case a difference in volume of some 10cc (or 3 points) can occur between measurers. A variation of a few points is therefore considered normal between measurers, and is significant only in borderline cases.

Several aspects of measurement rely on the integrity of the individual submitting the trophy. Measurements are therefore always made on trust and, in instances of deception, it is the deceiver not the measurer who is at fault, and instances of deception no doubt occur.

For the purposes of completeness, measurements of all species of the UK deer are included in the following chapters.

Equipment for trophy measuring

Steel tape: this should be of the 'engineer's' type, also called 'diameter tapes', which are flexible, narrow and flat, and designed with a short run-off beyond the zero mark. This ensures that circumference measurements are possible, as well as antler length measurements being made more easily. The tape should be able to measure up to at least 1m in increments of 0.1cm.

Spring balances: a fine precision balance weighing up to 1,000g in increments of 10g (5g can be readily estimated) will be required. It will need a hook on which to hang the trophy. Another, more robust, balance will also be required, capable of weighing up to 10kg in increments no more than 100g. A 'Pesola' precision spring balance will satisfy the former requirement; a Salter will suffice for the latter. Precision suspension electronic **digital balances**, such as 'Kern' (weighing in increments of 5g up to 5kg), are also available. Electronic Postal scales, although expensive, provide an even greater degree of accuracy.

Water container: for roe only. This must be large enough to comfortably accommodate a trophy of 30cm in length and at least 15cm in width.

Pencil/chalk: for marking antler at mid-distance (muntjac), tine eruption (red and sika), antler length and details of palm (fallow).

Electronic calculator.

15.

MEASURING RED STAG

This trophy can only be measured clean-boiled.

General notes on the measurement of red tines

These apply to all tines except the brows. Whereas sika tines are measured from the *outside*, red tines are measured from the *underside* edge of the tine. To establish the 'point of eruption' (see Figure 15.1), mark a chalk line (A) down the centre of the outside of the tine. Then mark another chalk line (B) down the centre line of the outside of the beam. Finally, mark a shorter line (C) towards the underside of the tine, which exactly bisects the angle created by lines (A) and (B). To be counted as a tine, a projection must be at least 2cm long. All measurements are taken in centimetres to the nearest 0.1cm.

Measuring and allocation of points

1. Length of antler

This measurement should be taken along the outside edge of the beam, from a point level with the underside of the coronet to the extremity of the longest tine in the crown. The tape should be allowed to fall naturally over the coronet onto the beam; it should not be pressed into the groove.

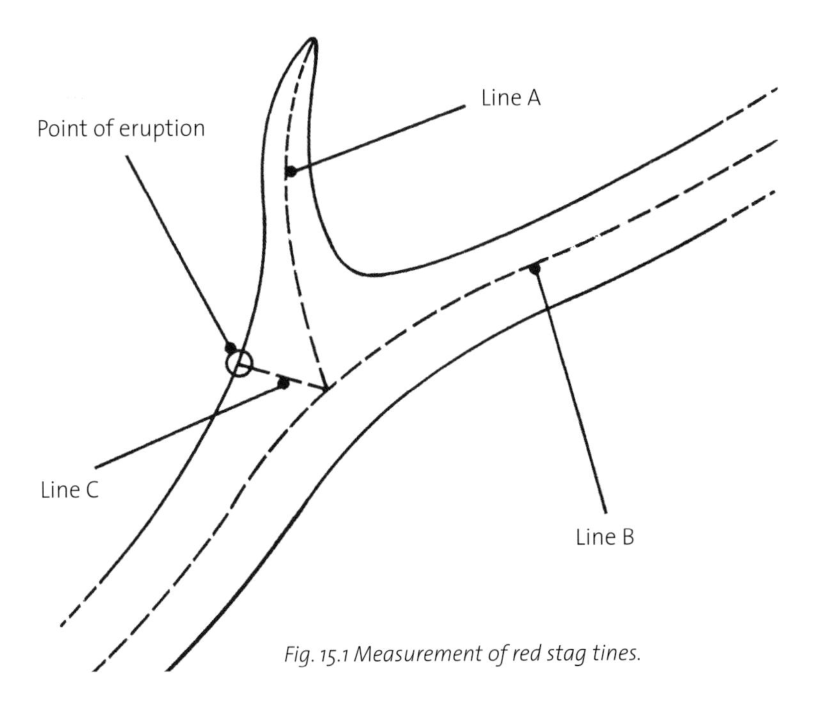

Fig. 15.1 Measurement of red stag tines.

Technique

It is almost impossible to measure a red trophy with accuracy without the aid of an assistant. Laying the trophy on its side on the floor, and using the assistant to steady it firmly, identify all the possible candidates for 'longest tine in the crown' and measure each one in turn. Record the greatest measurement as the length of each main beam, irrespective of its position within the crown. Determine the average of the two measurements, and multiply by a factor of 0.5 to establish the points scored for this section.

It is good practice at this stage with the larger species to continue to make all measurements on one antler at a time, rather than switching from left to right alternately.

2. Length of brow tine

Measure along the underside of the tine from the upper edge of the coronet to the extremity of the tine.

Technique

This measurement is made easier with the trophy turned upside down. Determine the average of the two measurements, and multiply by a factor of 0.25 to establish the points scored for this section.

3. Length of tray tine

Measure along the underside of the tine, from the point of eruption from the main beam to the tip.

Technique

Establish the point of eruption as instructed above, and mark with chalk before measuring. Determine the average of the two measurements, and multiply by a factor of 0.25 to establish the points scored for this section.

4. Circumference of coronet

Measure the circumference of each coronet.

Technique

Use your assistant to support the tape on the back of the coronet while the measurement is being taken. Determine the average of the two measurements to establish the points scored for this section.

5. Circumference of lower beam

This measurement should be taken at the narrowest place between the brow and the tray tine.

Technique

Loop the tape around the beam and move up and down until the lowest measurement is shown, including on either side of the bay. The measurement for each antler is carried forward as points scored for this section.

6. Circumference of upper beam

This measurement should be taken at the narrowest place between the tray and crown tines.

Technique

Loop the tape around the beam and move up and down until the lowest measurement is shown. The measurement for each antler is carried forward as points scored for this section.

7. Dry weight of antlers

This is taken to the nearest 10g. A deduction of 0.7kg should be made if the trophy has been prepared 'full skull'. For a long-nose cut, deduct 0.5kg.

Technique

Suspend the trophy from a suitable spring balance, and read off the weight. In the case of very large trophies, it may be difficult to support the trophy adequately to make an accurate reading. In this instance, the spring balance should itself be suspended from a firm surface, with no human interference. For reference purposes, a selection of skull off-cuts should be kept to establish a fair deduction in the case of intermediate cuts. Weight must not be added back in the case of very severe cuts. The dry net weight is multiplied by a factor of 2 to determine the points scored for this section.

8. Inside span

Measure at the widest point between the beams.

Technique

Place the tape *behind* the beams and move up and down and record the greatest reading. Score as follows:

Less than 60 per cent of average length of beam	0 points
From 60–69.9 per cent of average length of main beam	1 point
From 70–79.9 per cent of average length of main beam	2 points
80 per cent and over of average length of main beam	3 points

9. Number of points

A tine must be at least 2.0cm in length to qualify,

Technique

Count the total number of points on the trophy and carry this across as the points scored for this section.

10. Beauty points

These are allocated on the following basis.

a. Colour

Score in increments of 0.5 points as follows:

Light grey, yellow or artificially coloured	0 points
Grey to light brown	1 point
Dark brown to black	2 points

b. Pearling

Score in increments of 0.5 points as follows:

Smooth	0 points
Average	1 point
Good	2 points

c. Tine ends

Score in increments of 0.5 points as follows:

Blunt or porous	0 points
Pointed, but with brown-coloured tips	1 point
Pointed, with white tips	2 points

d. Bay tines

Score in increments of 0.5 points as follows:

	One side	*Both sides*
Short (2–10cm)	0	0.5
Medium (10–15cm)	0.5	1.0
Long 15+cm	1.0	2.0

e. Crown tines

Score in increments of 0.5 points as follows:

Tines that make up the two crowns		*Points scored*
5–7	short tines	1–2
5–7	medium tines	3–4
5–7	long tines	4–5
8–9	short tines	4–5

8–9	medium tines	5–6
8–9	long tines	6–7
10+	short tines	6–7
10+	medium tines	7–8
10+	long tines	9–10

Crown tines are defined as follows:

Short	2–10cm
Medium	10–15cm
Long	15+cm

Technique

Exercise sensible judgement where differences occur between left and right antlers, and apply average scores.

11. Penalty points

Up to 3 penalty points can be deducted for defects such as irregular attachment of antler to pedicle, pronounced lack of symmetry in length of beams, and any irregularity in brow bay or tray tines. Penalty points should not be applied to any irregularity that has already been taken into account during measurement of individual tines.

Technique

The concept is to make deductions for *deformities* associated with the tine, rather than with simple differences in length. A short tine will already have reduced the scoring potential of the trophy, whereas a misplaced tine would warrant a penalty at this stage.

Procedure for establishing overall points

Add up the points scored for each section, and then deduct any penalty points to establish the total points scored for the trophy.

CERTIFICATE OF MEASUREMENT
Red Stag

Owner's Name ....

Date Shot................................

Origin: Wild/Park...........................

Locality ....

Description			Measure	Total	Mean	Factor	Points
1. Length of Antler		L	cm	cm	cm	0.5	
		R	cm				
2. Length of Brow		L	cm	cm	cm	0.25	
		R	cm				
3. Length of Tray		L	cm	cm	cm	0.25	
		R	cm				
4. Circumference of Coronets		L	cm	cm	cm	1	
		R	cm				
5. Circumference of Lower Beams		L	cm			1	
		R	cm			1	
6. Circumference of Upper Beams		L	cm			1	
		R	cm			1	
7. Antlers	Weight		kg				
	Deduction		kg				
	Net Weight		kg			2	
8. Inside Span					%	0-3	
9. No. of Points						1	
10.	(a) Colour					0-2	
	(b) Pearling					0-2	
	(c) Tine Ends					0-2	
	(d) Bay Tines					0-2	
	(e) Crown Tines					0-10	
11. Sub-total							
12. Penalty Points						0-3	

FINAL SCORE [ ]

MEDAL AWARD LEVELS

Western European:	**Bronze 165**	**Silver 180**	**Gold 195+**
Scottish:	**Bronze 160**	**Silver 170**	**Gold 180+**

Medal Levels and Guidelines

Region	*Bronze*	*Silver*	*Gold*
Scottish	160–169.9	170–179.9	180+
Western European	165–179.9	180–194.9	195+
Eastern European:	170–189.9	190–209.9	210+

Guideline measurements

To achieve bronze medal (Western European), such as might be found in the low ground red deer populations of East Anglia, the New Forest and the West Country, a red trophy should have at least 12 points, and achieve the following measurements: antler length about 90cm; brow tine length about 30cm; tray tine length about 30cm; circumference of coronets about 24cm; upper and lower beam measurements about 12–13cm; dry net weight about 5kg. However, deficiencies in one area may be counteracted by increases in others, thus still ensuring that medal category is reached.

16.

MEASURING FALLOW BUCK

This trophy can only be measured clean-boiled.

General notes

Before starting to measure the trophy, use chalk to outline the basic shape of each palm (Figure 16.1 Line A) on its outer face. Any mass of antler lying outside that shape might qualify as a 'palmated speller' and thus be ineligible for the purposes of length measurement. A palmated speller is defined as a mass of antler lying outside the basic shape of the palm, where the width of that mass, measured at its widest point, is less than 50 per cent of the width of the palm itself. If the width is greater than 50 per cent, then it is considered as a normal extension of the palm, and may then be included for the purposes of length measurement. (Palmated spellers are not commonly found.)

Chalk should also be used to mark a line representing the total length of each antler. This line (B) commences on the outer side of the antler at the uppermost indentation of the palm (having previously considered whether palmated spellers are present or not), and continues down the palm parallel to the leading edge of the antler, and thence down the central line of the main beam to the base of the coronet. A further chalk line (C) should be drawn to represent the widest point of the palm. This line is drawn at 90 degrees to line B, and extends from the leading edge of the palm to the widest indentation at the back of the palm. The 'centre of the palm' for the purposes of span measurement is the point on the reverse (or inside) of the palm exactly behind the point at which lines B and C meet.

Measuring and allocation of points

1. Length of antler

This measurement should be taken to the nearest 0.1cm along the outside edge of the beam, from a point level with the underside of the coronet to the highest point of indentation of the palm, not to the highest point of the palm.

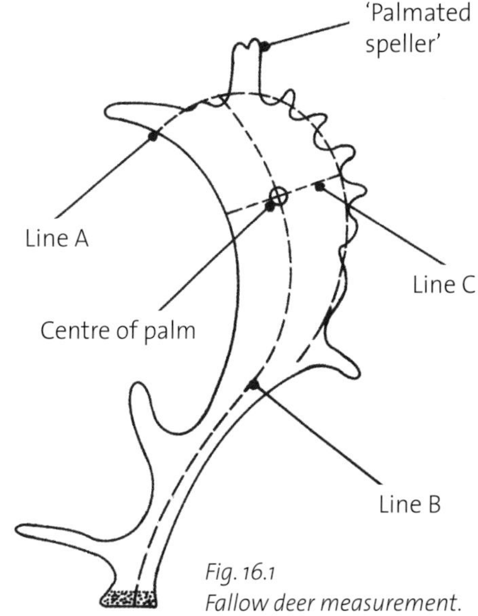

Fig. 16.1
Fallow deer measurement.

Technique

Mark the antler with chalk (line B as above), from the highest indentation of the palm parallel to the leading edge of the antler, down to the base of the coronet. Measure along that line, allowing the tape to fall naturally over the coronet onto the beam, and pressing it to the beam at a point 2cm above the coronet. Determine the average of the two measurements, and multiply by a factor of 0.5 to establish the points scored for this section.

It is good practice at this stage with the larger species to continue to make all measurements on one antler at a time, rather than switching from left to right alternately.

2. Length of brow tine

Measure along the underside of the tine, from the upper edge of the coronet to the extremity of the tine.

Technique

This measurement is made easier with the trophy turned upside down. Determine the average of the two measurements, and multiply by a factor of 0.25 to establish the points scored for this section.

3. Length of palmation

Measure parallel to the leading edge of the palm, from the point where it commences to broaden out, to the highest point of indentation.

Technique

Establish the point where the palms 'commence to broaden' by drawing a chalk mark at exactly 1cm above the narrowest point of the circumference of the upper beams. Measure along the outside of the palms up to the highest indentation. Determine the average of the two measurements to establish the points scored for this section.

4. Width of palmation

This is measured by taking the *circumference* measurement of the palm at the broadest place where an indentation appears on the back edge. This measurement must then be divided by two to determine the palm width.

Technique

The 'broadest place where an indentation appears' is determined by looping the tape around the palm to find the widest snag upon which the tape can be hooked. The measurement should follow line (C) from the widest indentation of the palm through a point at 90 degrees to line (B) and thence around the leading edge of the antler. Having divided the circumference measurement by two to get the palm width, this figure must be multiplied by a factor of 1.5 to establish the points scored for this section. The tape may be pressed into the palm when taking this measurement.

5. Circumference of coronet

Measure the circumference of each coronet.

Technique

Use your assistant to support the tape on the back of the coronet while the measurement is being taken. Determine the average of the two measurements to establish the points scored for this section.

6. Circumference of lower beam

This measurement should be taken at the narrowest place between the brow and the tray tine.

Photo opposite by Brian Phipps

Technique

Loop the tape around the beam and move up and down until the lowest measurement is shown, including on either side of the bay. The measurement for each antler is carried forward as points scored for this section.

7. Circumference of upper beam

This measurement should be taken at the narrowest place between the tray tine and the commencement of the palm.

Technique

Loop the tape around the beam and move up and down until the lowest measurement is shown. Upper beam measurement must never score more than 130 per cent of lower beam measurement. Any excess over that amount must therefore be ignored. The net measurement for each antler is carried forward as points scored for this section.

8. Dry weight of antlers

This is taken to the nearest 10g. A deduction of 0.25kg should be made if the trophy has been prepared 'full skull'. For a long nose cut, deduct 0.1kg.

Technique

Suspend the trophy from a suitable spring balance, and read off the weight. In the case of very large trophies, it may be difficult to support the trophy adequately to make an accurate reading. In this instance, the spring balance should itself be suspended from a firm surface, with no human interference. For reference purposes, a selection of skull off-cuts should be kept to establish a fair deduction in the case of intermediate cuts. Weight must not be added back in the case of very severe cuts. The dry net weight is multiplied by a factor of 2 to determine the points scored for this section.

9. Beauty points

These are allocated on the following basis.

a. Colour

Score as follows:

Pale yellow or artificially coloured	0 points
Grey or medium brown	1 point
Brown to black	2 points

b. Tine ends

Score as follows:	*One side*	*Both sides*
Short, few thin spellers	0	0
Spellers along one-third of palm edge	1	2
Spellers along two-thirds of palm edge	2	4
Spellers along entire length, plus back tine present	3	6

c. Mass, form, regularity

Up to 5 credit points may be awarded, of which 3 may be awarded for mass, and 2 for good shape and regularity.

Technique

Score 1 for mass between 2.75 and 3kg, 2 for mass between 3.01 and 3.25kg, and 3 for mass over 3.26 kg. Exercise sensible judgement where differences in regularity occur between left and right antlers, and apply average scores.

10. Penalty points

a. Insufficient span

Deduct points as follows:

Span less than 85 per cent of average length of beam	1 point
Span less than 80 per cent of average length of beam	2 points
Span less than 75 per cent of average length of beam	3 points
Span less than 70 per cent of average length of beam	4 points
Span less than 65 per cent of average length of beam	5 points
Span less than 60 per cent of average length of beam	6 points

CERTIFICATE OF MEASUREMENT
Fallow Buck

Owner's Name .... Date Shot...............................

Origin: Wild/Park.......................... Locality ....

Description			Measure	Total	Mean	Factor	Points
1. Length of Antler		L	cm	cm	cm	0.5	
		R	cm				
2. Length of Brow		L	cm	cm	cm	0.25	
		R	cm				
3. Length of Palmation		L	cm	cm	cm	1	
		R	cm				
4. Width of Palmation		L	cm	cm	cm	1.5	
		R	cm				
5. Circumference of Coronets		L	cm	cm	cm	1	
		R	cm				
6. Circumference of Lower Beams		L	cm			1	
		R	cm			1	
7. Circumference of Upper Beams (max score 130% of 6. above)		L	cm			1	
		R	cm			1	
8. Weight of Antlers	Weight		kg				
	Deduction		kg				
	Net Weight		kg			2	
9. Additions (Beauty)	(a) Colour					0-2	
	(b) Tine End					0-6	
	(c) Mass., Form, Regularity					0-5	
10.					Sub-total	1-9	
11. Penalty Points						Points	
(a) Insufficient Span	cm		%	1-6			
(b) Faulty Palmation				1-10			
(c) Unnatural Edges of Palmation				1-2			
(d) Irregular Antler				1-6			
12.					Sub-total		

FINAL SCORE [ ]

MEDAL AWARD LEVELS

Bronze 160 **Silver 170** **Gold 180**

Technique

Measure span at the widest point between the palms, and calculate that figure as a percentage of the average length of the beams (widest point measurement divided by average length, multiplied by 100).

b. Faulty palmation

Deduct points as follows:

	One side	*Both sides*
Swollen, diamond- or triangular-shaped palm	1–3	2–6
Bifurcated (split) palm	2–4	4–8
Jagged palm	3–5	6–10
Spike, or dagger-shaped palm	4–5	8–10

c. Unnatural edges of palmation

For smooth, worn or porous edge to palm 0–2 points can be deducted.

d. Irregularities

0–6 points can be deducted for antlers that are irregularly placed on the pedicles, or show differences between the length of beams, brows or trays.

Procedure for establishing overall points

Add up the points scored for each section, and then deduct any penalty points to establish the total points scored for the trophy.

Medal levels and guidelines

Bronze	*Silver*	*Gold*
160–169.9	170–179.9	180+

To achieve a bronze medal, a fallow trophy should achieve the following measurements: antler length about 60cm; brow tine length about 16cm; length of palmation about 30cm; width of palmation about 14cm; circumference of coronets about 20cm; upper and lower beam measurements about 12cm; dry net weight about 3kg. However, deficiencies in one area may be counteracted by increases in others, thus still ensuring that medal category is reached.

17.

MEASURING SIKA STAG

THIS TROPHY CAN be measured fully mounted or clean-boiled. Much controversy surrounds the subject of sika hybridisation. A useful guideline, based on nasal bone length (as measured from the skull suture junction to the tip of the nose bone), is as follows:

Length of nasal bone (cm)	*Species*
<8.5	Japanese
8.5–11.5	Manchurian, Formosan, etc. or hybrid
>11.5	Red hybrid

General note on the measurement of sika tines

This applies to all tines except brows. While red tines are measured from the *underside,* sika tines are measured from the *outer* edge of the tine (see Figure 17.1). To establish the point of eruption, lay the tape along the main beam of the antler, as if the tine were not present but following the natural mound caused by the base of the tine, and mark with chalk (A). Then place another chalk line (B) down the centre of the outer edge of the tine. The 'point of eruption' is the point at which these two lines meet. To be counted as a tine, a projection must be at least 2cm long, and its length must exceed the width at its base. All measurements are taken in centimetres to the nearest 0.1 of a centimetre.

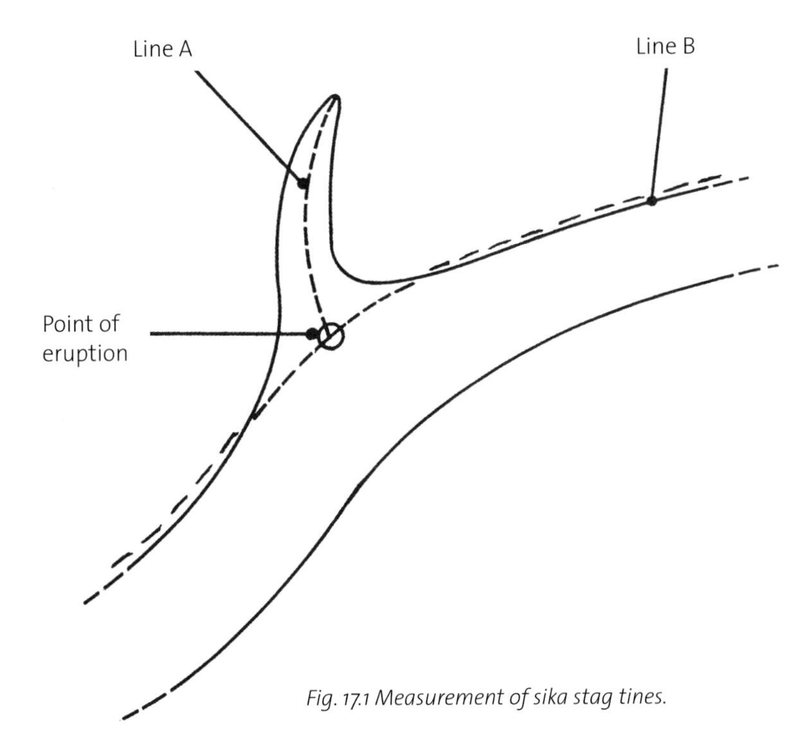

Fig. 17.1 Measurement of sika stag tines.

Measuring and allocation of points

1. Inside span

This measurement should be taken at the widest point between the normal second and third tines on the main beam. If the latter are absent, the measurement must be taken at the widest point, mid-distance between the second tine and the tip of the antler. Any span measurement in excess of the length of the longer antler must be entered in column D of the certificate of measurement as a deduction.

Technique

Sitting with trophy facing towards you, place the tape *behind* the beams and move up and down until the widest point at the perpendicular is found. A wide span is desirable, but it is important to check span against the length of the longer of the two main beams. If the span is indeed wider, deduct the longer beam measurement from the span, and the resulting figure will become the deduction in column D of the certificate of measurement.

2. Lengths of all abnormal tines

Calculate the sum of all abnormal tines placed between the brow and the third tine and enter as a deduction in column D of the certificate of measurement.

Technique

This measurement applies only to extra tines between the brow and third tine, and does not apply to extra crown tines, which are dealt with below. Note that only tines of at least 2cm, and of length greater than the width at their base, should be counted.

3. Length of main beam

Measure the outside curve of each antler from a point level with the lower edge of the coronet up to the tip of the antler, laying the tape naturally over the shoulder of the coronet and pressing at a point 2cm above it.

Technique

It is good practice at this stage with the larger species to continue to make all measurements on one antler at a time, rather than switching from left to right alternately.

4. Length of brow tine

Measure from the upper edge of the coronet along the lower edge of the tine to its tip. If the brow tine emanates from the beam at a point 5cm or more above the coronet, brow length will be measured from the point of eruption, as defined in the General Note at the start of this chapter.

Technique

This measurement is made easier with the trophy turned upside down.

5. Length of second tine

Measure along the outer edge of the tine, from the point of eruption from the main beam to the tip.

Technique

Establish the point of eruption as instructed above, and mark with a pencil before measuring.

6. Length of third (inner) tine

Measure along the outer edge of the tine, from the point of eruption from the main beam to the tip.

Technique

Establish the point of eruption as instructed above, and mark with a pencil before measuring. A representative Japanese sika stag should have a full complement of eight (four+four) tines, but six-pointers (three+three) are quite common and will simply forego any additional marks associated with this section. Note that only tines of at least 2cm, and of length greater than the width at their base, should be counted.

7. Length of first extra crown tine

Measure as in 5 and 6 above.

Technique

Extra crown tines are acceptable, but rarely present as pairs in Japanese sika. An extra crown tine on one beam only will form an addition in column B or C of the certificate of measurement, but then be cancelled out by a deduction in column D. Extra crown tines are much more common in Manchurian sika.

8. Length of second and third extra crown tines

The measurement of these tines is performed as in 5 and 6 above, applying the same technique as in 7 above.

9. Circumference of lower beam

This measurement should be taken at the narrowest place between the brow and the second tine.

Technique

Loop the tape around the beam and move up and down until the lowest measurement is shown.

10. Circumference of upper beam

This measurement should be taken at the narrowest place between the second and inner tine. If the inner tine is not present, then the measurement should be taken halfway between the middle tine and the tip of the main beam.

Technique

Loop the tape around the beam and move up and down until the lowest measurement is shown.

Procedure for establishing overall points

The length and circumference measurements in centimetres are added up for each antler and tine, and then added to the inside span measurement. Any differences in measurement (column D of the certificate of measurement) between right and left antler are also totalled to become a deduction before establishing a final score.

Medal levels and guidelines

Type	*Bronze*	*Silver*	*Gold*
Japanese	225–239.9	240–254.9	255+
Manchurian, Formosan, etc.	300–349.9	350–399.9	400+

To achieve a bronze medal, a Japanese sika should be a fairly even 8-pointer with a span of about 40cm, beams of about 50cm, brows of about 13cm, second tines of about 14cm and inner tines of about 6cm. The circumference of the lower beams should be about 9cm and the upper about 6.5cm. However, deficiencies in one area may be counteracted by increases in others, thus still ensuring that medal category is reached.

CERTIFICATE OF MEASUREMENT
Sika Stag

Owner's Name .... Sub-Species............................

Origin: Wild/Park.......................... Date Shot..............................

Locality ....

Number of Tines (left)			Number of Tines (right)		
Tip to Tip (cm)			Greatest Spread (cm)		
		A	B	C	D
SCORE DATA		Span Credit (1)	Left (cm)	Right (cm)	Difference (cm)
1. Inside Span between Main Beams					
2. Total lengths of all abnormal Tines Between Brow & Third Tine					
3. Length of Main Beam					
4. Length of Brow Tine					
5. Length of Second Tine					
6. Length of Inner Tine					
7. Length of first Extra Crown Tine					
8. Length of second Extra Crown Tine					
9. Length of third Extra Crown Tine					
10. Circumference of Lower Beam					
11. Circumference of Upper Beam					
12. Column Sub-totals					
		Add A+B+C	Subtract D=	FINAL SCORE	

MEDAL AWARD LEVELS

Japanese:	**Bronze 225**	**Silver 240**	**Gold 255**
Manchurian etc.:	**Bronze 300**	**Silver 350**	**Gold 400**

18.

MEASURING ROE BUCK

THIS TROPHY CAN only be measured in a clean-boiled state with all residual velvet removed.

Measuring and allocation of points

1. Length of antler

Each antler is measured to the nearest 0.1cm along the outside edge of its beam, from the lower edge of the coronet to the tip of the top tine. The tape should be allowed to fall naturally over the coronet onto the beam, and then be pressed to the beam at a point 2cm above the coronet. The average length of the two antlers is then calculated, and that average divided by 2 to give the length score.

Technique

(As for a right-hander.) Lay the trophy on its side, either on your lap or on a desk, with the top tine facing towards your right. Grip the tape between left forefinger and thumb, with the zero mark level with the lower edge of the coronet. With a firm grip on the left hand, extend the tape up the outside edge of the antler, if necessary gripping the tape with the forefinger and thumb of the *right* hand, and bringing the left hand up to replace that grip, at any point where a change of direction occurs. The measurement should be taken with the minimum number of changes of direction, and should represent as straight a line as possible up the beam. It is a useful tip to complete the appropriate section of Section 4 (Inside Span) as soon as the average length has been calculated.

2. Weight of antlers

The gross weight in grams (to at least the nearest 5g) is recorded, with a deduction made according to the way that the trophy has been cut. This net weight is then divided by 10 to give the weight score.

Technique

Suspend the trophy from the spring balance, antlers towards the floor, with an elastic band placed around the skull behind the pedicles. Support the *suspending arm* with a hand at its elbow, with the *supporting arm* placed on a firm surface at a level to ensure ease of spring balance reading. The net weight required is the equivalent to a skull that has been 'standard' or 'short-nose' cut. For a 'full skull' (i.e. retaining all skull bones and the upper teeth), a standard deduction of 90g is made. For a 'long-nose cut' (i.e. retaining the entire eye socket, the jaw anchorage bone and the entire nasal cavity, and with just the upper jaw removed) the standard deduction is 65g. For other variations of cut an appropriate deduction must be made, and this is made easier if a collection of off-cuts is retained for this purpose. Note that weight may not be added back in the case of a trophy cut shorter than standard cut. As soon as the *gross weight* is entered in the weight section of the certificate of measurement, it is a useful tip to immediately enter the 'Weight in Air' – Section 3 (Volume).

3. Volume of antlers

The volume in cubic centimetres is calculated by deducting the weight in grams of the trophy with antlers immersed in water, from the dry gross weight in grams of the entire trophy. This volume is then multiplied by 0.3 to obtain a volume score. When immersing the trophy, no part of the skull or pedicles should be immersed.

Technique

This is the measurement most likely to cause difficulties, particularly as re-measurements of a recently immersed trophy result in misreads. It is therefore important to prepare properly for this point of measurement. Fill a large container (comfortably bigger than the trophy itself) with water to as near the top as possible. Place the container in a well-lit area, at a level which makes reading the scale as easy as possible. Using a standard suspension balance, support the trophy as in Section 2 above, check the dry weight and record before slowly but positively immersing in water up to as much of the coronet as is possible, whilst resting the forearm on the edge of the container for support.

Remembering that no part of the skull or pedicle should be immersed, the elastic band should be eased back or forward to ensure the best point of balance for the trophy before immersion. However, in the case of trophies with 'roof coronets', it is permissible to immerse the trophy to a point which averages any pedicle which is immersed against any antler which remains out of the water. If an over-immersion occurs, re-measurement should not take place until the trophy is allowed to dry completely. If using an electronic digital suspension balance, suspend the trophy from the hook, recalibrate to zero and dip as above. The resultant reading will be the volume in cubic centimetres. A useful, but by no means infallible, check on the accuracy of your dip is to bear in mind that in a *normal* trophy, the *weight score* is usually about 80 per cent of the volume score. This percentage tends to rise with the age of the buck, but is very rarely greater than the volume score.

4. Inside span

Measure the inside span at the widest point perpendicular between the main beams. Calculate the span as a percentage of the average beam length (i.e. divide the span by the average beam length and multiply by 100), and score as follows:

Percentage	*Points*
Less than 30	0
30–34.9	1
35–39.9	2
40–44.9	3
45–75	4
Over 75	0

Technique

Sitting with the trophy on your lap facing towards you, place the tape *behind* the beams and move up and down until the widest point at the perpendicular is found. Holding the trophy at arm's length will ensure the most appropriate measurement. In the case of 'racked' antlers, the measurement should be taken at a point perpendicular in both planes. In the case of the antlers being of unequal length, and if the widest point is at the tip of one of the top tines, then the measurement should be taken perpendicular to the tip of the shorter tine. The concept of span is that it is represented by the shadow which would be formed on a flat surface by shining a bright light from directly in front of the trophy.

5. Colour

Score as follows in increments of 0.5 of a point:

Pale or artificially coloured	0 points
Yellow/grey or pale brown	1 point
Light brown	2 points
Dark brown	3 points
Black	4 points

Technique

The rules are very clear and the trophy should be thoroughly examined on all sides. Where velvet exists, that part of the antler is deemed to be white. An average of the entire trophy should be taken into account when giving a score, ignoring polished tips that are quite normally white. *Any* artificial colouring disqualifies the trophy, even if only added to fill colour where, for example, velvet has been removed. Artificial colour is often exposed during the volume dip, and specks may be visible on the upper part of the skull.

6. Pearling

Score as follows in increments of 0.5 of a point:

Smooth, almost no pearling	0 points
Weak pearling	1 point
Average pearling (small and sufficiently numerous)	2 points

Good pearling (small but on both sides of beam)	3 points
Very good pearling (well pearled throughout all parts of beam)	4 points

Technique

It is good practice to measure as if against perfection. Sometimes trophies are presented with full pearling from base to tip of both sides of beams, and with pearling on all tines. This will score the full 4 points. Note that while it is common (i.e. 'average') that pearling is found on the *inside* of the beams, 'good' pearling will be represented by additional pearling on the *outside* of the beam. Differences occur between trophies exhibiting large pearls as against small pearls. Both can score the maximum, the crucial factor being the percentage cover over the entire antler.

7. Coronets

Score as follows in increments of 0.5 of a point:

Feeble or weak (thin and flat)	0 points
Average (as a girdle with small pearling)	1 point
Good (in the form of a crown, sufficiently high)	2 points
Strong (large and high)	4 points

Technique

Again, it can help to measure as if against perfection. Trophies are occasionally presented with huge coronets (4–5cm across and 2cm deep), which very obviously score the full 4 points. However, difficulties occur with roof coronets, and where the coronet is well developed, but poorly defined against the beam. The important factor is to assess the entire girdle of the coronet, particularly at the back of the antler. Differing styles of coronet can thus achieve the full 4 points, but only if their development is to standard. It can be useful to judge the coronet in proportion to the beam.

8. Tine ends

Score as follows in increments of 0.5 of a point:

Blunt or poorly developed	0 points
Blunt but of medium development	1 point
Pointed and white-tipped	2 points

Technique

If a missing tine is penalised at this point, then it should not be penalised once again under Section 10 below. Broken tines are ignored, and assessment is made only of the remaining unbroken tines.

9. Additional points for regularity and quality

Score in increments of 0.5 of a point. Up to 5 points in total can be awarded, of which 3 points may be reserved for regularity of tines and up to 2 points for their quality scored as follows:

Normal tines	0 points
Good tines	1 point
Very good tines	2 points

Technique

The concept is that points should only be awarded for regularity or quality that is above the norm. The difficulty is in grading regularity. Nevertheless it is useful to score as against perfection.

10. Penalty Points

Score in increments of 0.5 of a point. Up to 5 points may be deducted in total, of which 3 may relate to various unspecified abnormalities, and up to 2 for the following tine defects:

Mediocre tines or emanating from one beam only	1 point
Tines absent or very short	2 points

Technique

At this point, broken tines may be penalised, but account should be taken of any score attributed to Section 8 above, as a deduction must not be made for the same reason twice. In the event of a severely malformed trophy, it is useful to consider each antler separately as being eligible for 2.5 deductions on its own account. Additional tines may not qualify for deduction if evenly placed on both antlers, but tines or jags present on only one antler should be considered.

Procedure for establishing overall points

It is good practice to award all the 'beauty points' (Sections 5–10), *before* making the physical measurements (Sections 1–4). This ensures that there is absolutely no possible chance of being tempted to manipulate the final score in any way. In any case, the beauty points should be assessed before the head is immersed in water.

The total of all section scores gives a final score.

Medal levels and guidelines

Bronze	*Silver*	*Gold*
105–114.9	115–129.9	130+

The most important facet of a roe trophy is its weight in grams, which accounts for about 35 per cent of the entire score. Because many trophies are cut to individual preferences, home mis-measurements are often made. In effect, we are seeking the net weight of the trophy as if standard cut, which is from the highest point of the dome of the skull, through the eye sockets and midway through the nasal bone. Volume accounts for a further 45 per cent of the total score, and of the remaining 20 per cent, about half is in the average length of antler (ideally between 20 and 30cm) and remainder in the beauty points.

A bronze medal will generally need a dry net weight of at least 365g (455g full-skull) and a volume of at least 150cc. A silver will probably require a dry net weight of at least 420g (510g full-skull) and a volume of at least 165cc, and a gold normally requires a weight of 480g (570g full-skull) and a volume of 200cc. However there is considerable variation arising mainly to the age of the buck and also to the quality of beauty points. Trophies of 450g net have been known to make gold, and trophies of over 400g sometimes fail to make even bronze. Nevertheless, a trophy of 26cm length, of 365g net weight and 150cc in volume, scoring full marks for span and average marks for each of the beauty points, will score a total of 105.5 points – bronze.

Photo opposite by Brian Phipps

CERTIFICATE OF MEASUREMENT
Roebuck

Owner's Name ....

Number of Points .... Date Shot............................

Approximate Age .... Locality ....

			Points
1. Length of Antler	L cm R cm Total cm	Av. cm	0.5	
2. Weight	grms Deduct grms Net Weight grms	0.1	
3. Volume	Weight in Air grms Weight in Water grms Volume cm^3	0.3	
4. Inside Span Span / Av. Length	cm / cm x 100 =	%	
5. Colour	(max. points = 4)		
6. Pearling	(max. points = 4)		
7. Coronets	(max. points = 4)		
8. Tine Ends	(max. points = 2)		
9. Regularity and Quality	(max. points = 5)		
		Total for 1 to 9	
10. Penalty Points		(max. points = 5)	
		FINAL SCORE	

MEDAL AWARD LEVELS

Bronze 105 **Silver 115** **Gold 130**

19.

MEASURING MUNTJAC BUCK

This trophy can be measured fully mounted or clean-boiled.

Measuring and allocation of points

1. Inside span

This measurement should be taken at right angles to the centre line of the skull at the widest perpendicular point between the main beams.

Technique

Sitting with the trophy on your lap facing towards you, place the tape *behind* the beams and move up and down until the widest point at the perpendicular is found. Holding the trophy at arm's length will ensure the most appropriate measurement.

2. Length of main beam

Measurement should be taken along the outside of the antler from a point level with the bottom edge of the coronet up to the tip of the antler, passing the tape naturally over the shoulder of the coronet and pressing it in at a point 2cm above the coronet.

Technique

Measure up the outside of the antler to the point of any pronounced curve commonly found towards the top of the antler, and continue over the shoulder of this curve until the extremity of the tip. Whilst taking this measurement, a useful tip is to mark with pencil the halfway point to assist with Section 5 below. In the case of very sloped coronets, the measurement should be taken from a point at the lower edge of the coronet, and midway through the pedicle as seen with the skull laid on its side.

3. Length of brow tine

This measurement should be taken from the upper edge of the coronet, along the underside of the tine, up to its tip.

Technique

Note that a brow will only count as a tine if it is at least 1cm long, *and* if its length exceeds the width at its base. It is therefore commonly the case that although a projection has the appearance of a tine, it may technically fail these tests and not therefore count.

4. Circumference of coronet

Measure the circumference of each coronet.

Technique

Although upright coronets present few problems, in the case of severely sloped coronets it is useful to have someone to hold the tape at the back of the coronet while the measurement is taken.

5. Circumference of beam

Measurement is taken at the mid-distance point up each beam.

Technique

The mid-distance point is readily marked while each antler is being measured as in Section 2 above. If they have not already been marked, divide the length of each antler by 2, and mark their respective mid-distances with a pencil.

Procedure for establishing overall points

The length and circumference measurements in centimetres are added up for each antler and brow tine, and then added to the inside span measurement. Any differences in measurement between right and left antler are also totalled to become a deduction before establishing a final score.

Medal levels and guidelines

Bronze	*Silver*	*Gold*
56–58	58.5–60.9	61+

Guideline measurements

A gold medal trophy will need to be an even head of about 12cm in length, in excess of 10cm width between the beams, and with brow points in excess of 1cm. The circumference of the coronets should be about 8cm, suggesting a mid-antler length circumference of about 4cm. A trophy can still attain medal category without brow tines, but only if compensated by commensurate increases in length, span or circumference measurements. Evenness is crucial, as the effect of an uneven trophy is that the overall score becomes relative to the smaller antler.

Photo by Brian Phipps

CERTIFICATE OF MEASUREMENT
Muntjac Buck

Owner's Name .... Date Shot..............................

Approximate Age .... Locality ....

Number of Tines (left)		Number of Tines (right)	
Tip to Tip	cm	Greatest Spread	cm

SCORE DATA	A Span Credit (cm)	B Left (cm)	C Right (cm)	D Difference (cm)
1. Inside Span between Main Beams				
2. Length of Main Beam				
3. Length of Brow Tine				
4. Circumference of Coronet				
5. Circumference of Antler at mid- distance up beam				
6. Column Totals				
	Add A+B+C	Subtract D=	**FINAL SCORE**	

MEDAL AWARD LEVELS

Bronze 56 **Silver 58.5** **Gold 61**

20.

MEASURING CHINESE WATER DEER BUCK

This trophy can only be measured with canines removed.

Measuring and allocation of points

1. Length of canine tooth

Each canine should be measured in millimetres along the front curve (leading edge) of the tooth, from the base of the canine to its tip.

2. Circumference of canine at point of eruption

The circumference should be measured in millimetres as close as possible to the point of eruption from the jaw.

Technique

This measurement is extremely difficult to take using a steel tape, and takes years off the life of even the best quality tapes. As the tooth has been removed for the purposes of measurement, then the measurer will probably require the skull to ensure that the point of eruption can be established. Sometimes there is a clear colouration and surface difference which delineates the point of eruption. If not, then the tooth should be replaced until it meets the point of resistance, and the circumference measurement then taken. A useful tip is to replace the tooth, mark the

CERTIFICATE OF MEASUREMENT
Chinese Water Deer Buck

Owner's Name .... Date Shot................................

Origin: Wild/Park.......................... Locality..................................

	1	2	3
	Left	Right	Difference
1. Length of Canine	mm	mm	mm
2. Circumference	mm	mm	mm
3. Totals			
4.	Col. 1 + 2	Subtract Col. 3	**Final Score**

MEDAL AWARD LEVELS

Bronze 180 **Silver 190** **Gold 210**

point of eruption with a pencil, and then remove it again for ease of measurement.

Procedure for establishing overall points

The length and circumference measurements in millimetres are added up for each canine and then added together. Any differences in measurement between right and left canine are also totalled to become a deduction before establishing a final score.

Medal levels and guidelines

Bronze	Silver	Gold
180–189.9	190–209.9	210+

Guideline measurements

A gold medal will require canines of an even length of at least 75mm, with a circumference at the point of eruption of at least 30mm.

21.

TRAVELS IN KURGAN PROVINCE AFTER THE SIBERIAN ROE

TO ANY ROE ENTHUSIAST the chance to study and even shoot the Siberian roe must be the ultimate dream. For me that dream came true in 2007 at the kind invitation of Monsieur Arnaud Brunel.

I knew that our destination was the foothills of the Urals, but I had no idea of the scale of the place. These foothills extend to beyond 300km of the actual mountains and the topography is flat, very flat. There are huge agricultural fields, some over several hundred hectares in size, and birch forest with a few scrubby willow. No other trees, just birch, and the occasional clearing or larger grass field, cut for hay, much of which is loosely stacked up to nearly 5m high to feed the roe and the moose during the long winter with up to six months of snow cover. The black soil could grow anything but here, so recently transformed to agribusiness after years of small-scale collective farming, the main crops were wheat, barley, oats and rye. There were also huge areas of lucerne set aside purely for the roe. It was autumn, the sun shone, the colours were magnificent and for the main part it was warm and dry. Snow would not come for a few weeks yet.

But even getting there, armed, alone and in the middle of the night, presented its challenges. I left the moderator behind – illegal in much of Europe – I could only imagine what the Russian authorities might make of it. Touching down at 2.00am in Yekaterinburg, site of the infamous slaughter of the Romanovs in 1918, to the surly frowns of the immigration

ABOVE *Birch forest with a few scrubby willows*

LEFT *No other trees, just birch, and the occasional clearing or larger grass field, cut for hay much of which is loosely stacked up to 5m high to feed the roe and the moose during the long winter of up to six months snow cover.*

officials was intimidating, but transition was smoothed by the appearance of our translator, and the many and various forms had soon received their official, and very heavy, stamp.

After a four-hour minivan drive we reached our destination – simple accommodation by western standards but five star to the locals. There were twin log cabins for sleeping, a dining hut, a sauna for washing (there being no mains water system in that area), and a very basic outside privy. While the new Russian middle-class city dwellers take full advantage of the new democracy and open trade, the country dwellers remain locked in the nineteenth century. We visited a local village and encountered people (all elderly) who had never met a foreigner. Their houses were single-storey wooden cabins, shabby now, but displaying ornate carvings which suggested better times. All their churches had been burned by the Communists, and the remaining shells were haunting. Their vehicles were all from the 1950s or early '60s and the main garden crop was cabbages. Despite all this, the people who looked after us were magnificent and proud: knowledgeable guides, a good cook and the ever-present translator. We had four square meals a day and plenty of vodka (drunk like wine).

On my first evening there we must have seen 200 roe, clustered into loose groups of up to 30, mainly on the lucerne fields, but also on the stubbles. The Siberian roe is found anywhere between the Urals in the

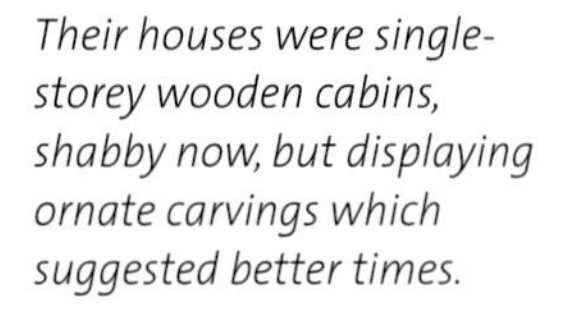

Their houses were single-storey wooden cabins, shabby now, but displaying ornate carvings which suggested better times.

OPPOSITE PAGE *All their churches had been burned by the Communists, and the remaining shells were haunting.*

Their vehicles were all from the 1950s or early '60s.

west and the Altai Mountains in the east, with the Kurgan region being known for holding the highest density. These deer are similar to the European roe, but bigger. They are much more vocal and their bark is rather eerie, the nose is more Roman, the back flatter, the winter coat paler and the antlers typically much bigger, wider and of eight points. Nevertheless, they are completely recognisable and the size difference is not instantly apparent. Twins are the norm, with triplets being common – clearly they do very well in this environment in the absence of any remaining natural predators. But what was so surprising and captivating was the wealth of other game and wildlife – capercaillie, black grouse, grey partridge, quail, woodcock, duck, geese, eagles, harriers, massive wild boar, moose, and something called a possum dog. I suppose that western Europe must have looked like this once upon a time.

The deer were mostly seen in the lucerne fields.

Siberian roe are similar to European roe, but bigger; their winter coat paler and the antlers typically much bigger.

Nevertheless they are completely recognisable.

Sadly, stalking was done mostly from vehicles and the roe were unsurprisingly completely car-shy (a lesson here for the law-makers who have recently made shooting from cars legal – it takes five years to tame your roe to vehicles and just one year to ruin it!). However, I managed to get one of the guides to take me foot-stalking and on that outing we encountered a magnificent buck, the vision of which I still carry more intensely than any trophy. We saw many roe and also several moose, mainly cows with their calves. We used some proper high seats, and later used the haystacks as high seats, and it was on one such outing that I was privileged to shoot a typical old buck, with thick, pale, yellow-grey winter coat and long, branched eight-point antlers – so familiar but so different, as big as an average sika but softer-skinned and falling instantly to the little 70 grain ballistic tip 243. Having shot one, I was now determined to get a good photograph. The other members of the party were all keen to shoot a moose, but it was me that the moose decided to grace with its presence. This was an extraordinary and memorable encounter – at 50m for ten minutes – with time enough to take both stills and video. It was an enormous and graceful animal which seemed pretty unruffled by our presence. Later on, a male capercaillie arrogantly blocked our path, displaying and attempting (with some success) to intimidate. We had many encounters with bucks but few opportunities for the perfect photograph. Eventually I had some success but, as ever, it proved much more difficult than shooting one with a rifle.

The recollections remain fresh today, and my abiding memory is of huge unspoilt landscape, of abundant and varied wild game, and of great companionship.

We used some high seats.

ABOVE *It was an enormous and graceful animal, which seemed pretty unruffled by our presence.*
BELOW *The recollections remain fresh today, and my abiding memory is of huge unspoilt landscape, of abundant and varied wild game, and of great companionship.*

From left: The author, Prince Stanislas Poniatowski, our host Arnaud Brunel, Comte Edouard de Boisgelin, our translator, the cook, the head stalker and the assistant stalkers.

22.

VENISON RECIPES
Marco Pierre White
Wheelers of St James

Fillet or rump of roe deer with wild mushrooms and Madeira roasting juices

Serves 2

INGREDIENTS

Extra virgin olive oil (or clarified butter)
Fillet or rump of roe deer, about 350g
100ml Madeira
150g wild mushrooms
Fresh herbs of your choice, to garnish

METHOD

1. Preheat the oven to 180 °C /gas 4.
2. Heat the olive oil or butter in a non-stick frying pan and caramelise the meat on both sides. Transfer to a roasting tin, skin-side down, place in the oven and roast for 10–15 minutes.
3. Remove the meat from the oven and allow to rest for 5–10 minutes.
4. Meanwhile, boil the Madeira in a saucepan until evaporated and reduced almost to syrup. Combine the roasting juices with the Madeira reduction.
5. In a pan, fry the mushrooms in olive oil over a medium heat for a minute or until they are cooked to your taste.
6. To serve, slice the roe and spoon the mushrooms over and around the meat. Pour over the syrupy juices and scatter with fresh herbs.

Fillet of roe deer with fresh ceps, vintage balsamico, olive oil and crystal salt

Serves 4

INGREDIENTS

250g of wild mushrooms (I have used porcini)
1 tablespoon balsamic vinegar
Extra virgin olive oil
4 fillets of roe deer steaks, about 180g each
Fresh herbs of your choice to garnish

METHOD

1. Clean and slice the mushrooms. Make a dressing by mixing the vinegar with 2 tablespoons of olive oil. Set aside.
2. In a heavy-based frying pan, heat a tablespoon of olive oil and fry the roe steaks for 3–4 minutes on each side, turning only once. Remove the pan from the heat and allow the roe steaks to rest in the pan in a warm part of the kitchen for 5–10 minutes.
3. Meanwhile, in another pan, fry the mushrooms in a generous amount of olive oil for a minute or two.
4. Using a pastry brush, paint the meat with the balsamic vinegar oil dressing and scatter the wild mushrooms on and around the roe steaks. Mix the juices from the mushroom pan with the juices from the pan used to cook the steaks, and spoon this lovely sauce over the meat. Scatter with herbs.

Fillet of roe deer with black pepper, raisins sec a la Armagnac

Serves 2

INGREDIENTS

2 dessert spoon raisins
2 fillet of roe deer 180g each
½ dessert spoon of cracked black pepper
1 tablespoon of extra virgin olive oil
Knob of unsalted butter
1 dessert spoon of Armagnac
Fresh herbs of your choice to garnish

METHOD

1. Put the raisins in a small saucepan, cover with cold water and bring to the boil. Reduce the heat and let the water simmer for about 5 minutes.
2. Remove the pan from the heat, drain the raisins and rinse them under cold water; set aside.
3. Put the cracked black pepper on the plate and dust the roe fillets.
4. Heat the olive oil in a heavy-based frying pan and fry the roe fillets for 3–4 minutes on each side, turning them only once.
5. Remove the pan from the heat and allow the roe fillets to rest in the pan in a warm part of the kitchen for 5–10 minutes.
6. While the roe fillets are resting, put the raisins in a hot pan with a knob of butter and cook for a minute or two, before adding the Armagnac and cooking for another 20 or 30 seconds. Serve with raisins on the top of and around the roe fillets and scatter the fresh herbs.

Grilled venison sausages with braised red cabbage

Serves 4

INGREDIENTS

12 best quality venison sausages
½ a red cabbage
250ml white wine vinegar
250g caster sugar
Salt and freshly ground black pepper

METHOD

1. Place sausages in a hot frying pan and cook until golden on all sides.
2. Cut the cabbage in half and remove the solid root. Carefully shred the leaves into fine ribbons.
3. In a large stainless steel pan, heat the vinegar. Add the sugar and, once it has dissolved, add the shredded cabbage. Stir to cover all the cabbage in the vinegar. Turn the heat down to its lowest setting and place a lid on the pan. Slowly braise the cabbage for about 1 hour until it is tender and the vinegar has reduced to syrup. Adjust the seasoning as necessary. If the cabbage is a little 'sharp', add a touch more sugar. (This cabbage keeps for about a week when stored in a fridge.)

Wheelers of St James roast haunch of venison, sauce grande veneur

Serves 4–6

INGREDIENTS

1 haunch on the bone
600g venison bones, chopped into 2cm pieces
400g venison trimmings
50ml groundnut oil
35g fresh butter
100g mirepoix (see below)
80ml cabernet sauvignon vinegar
600ml red wine reduced to 400ml
1.2 litres brown chicken stock
200g button mushrooms, sliced
1g juniper berries
10ml red wine vinegar, reduced by half
20ml whipping cream
25ml cognac, boiled for 10 seconds
40g redcurrant jelly
15g bitter chocolate (100 per cent cocoa)

METHOD

The mirepoix

This is made using the following vegetables, all diced into 2cm cubes:

150g onion
75g carrot
36g leek
36g celery

1. Peel onions and dice.
2. Peel carrots and dice.
3. Leeks – peel first layer and wash, then dice.
4. Celery – take off stalk head, then wash and dice.

Once all vegetables are chopped, mix well so that when used in stocks and sauces there is an even distribution of vegetable in the mirepoix.

The sauce

1. Caramelise bones and trimmings dark brown, first in oil then in butter for 30 minutes: reserve one-third to refresh.
2. Drain.
3. Caramelise the mirepoix in foaming butter; drain.
4. Mix bones and mirepoix.
5. Deglaze the roasting dish with cabernet sauvignon vinegar and reduce down.
6. Add the red wine, reduce by two-thirds, then add the brown chicken stock.
7. Bring to the boil, skim and simmer for 1 hour.
8. Strain, pass through a fine chinoise.
9. Bring to the boil.
10. Add mushrooms, juniper berries, red wine vinegar reduction, cream and trimmings; refresh.
11. Pass through a sieve.
12. Finish with cognac, redcurrant jelly.
13. Taste and correct seasoning.
14. During service finish the sauce with 1g per portion bitter chocolate.

The meat

1. Season and caramelise the haunch on all sides in a hot roasting pan.
2. Roast at 160 °C for approximately 60 minutes.
3. Remove to a meat dish and rest in a warm place for 15–20 minutes.

Wheelers of St James venison pie

Serves 4

INGREDIENTS

250g onions, chopped roughly
200g leeks, chopped roughly
200g carrots, chopped roughly
200g celery, chopped roughly
500g venison stewing meat
2 litres of brown chicken stock
1 bottle of red wine
2g whole juniper berries
1 litre double cream
1 bunch of thyme
50g redcurrant jelly

METHOD

1. Cut the venison meat into 4cm cubes, cover in flour and dust off excess.
2. Seal in a frying pan and colour. Once all the meat is a deep brown colour all over, remove from the pan, pour off excess fat and deglaze with red wine. Reduce red wine by half.
3. Colour all the vegetables in a little oil and add to the venison and wine. Meanwhile, bring the chicken stock to the boil and add all other ingredients; bring back to the boil and skim. Season with salt and pepper, place in a casserole, cover and place in a low oven for 6 hours or until the meat is tender.
4. Allow to cool and then strain through a colander; remove the vegetables and retain the meat.
5. Reduce the braising liquid until it forms a sauce consistency, skimming it as it reduces.
6. Place the meat back into the sauce and season once more.

Use as desired; you can either use it as is as a stew or place in a pie dish and cover with puff pastry.

Wheelers of St James accompaniments for roast saddle of roe deer

Wheelers of St James roasted chestnuts

INGREDIENTS

750g gross (500g net) chestnuts, peeled
120g gross (70g net) onions, chopped
70g gross (30g net) celery, chopped
50g butter
5g salt
15g sugar
1g finely ground white pepper
200ml white chicken stock
20ml lemon juice
2g thyme

METHOD

1. Sweat onion and celery in butter for 3–4 minutes without colour.
2. Add salt, sugar, pepper, white chicken stock, lemon juice and thyme.
3. Put the chestnuts into the pan and simmer gently at 89–90°C for approximately 40 minutes.
4. Cool, in bags, on a tray and refrigerate until required.

Wheelers of St James creamy cabbage

INGREDIENTS

½ a savoy cabbage
100g onion, diced
1 clove garlic
400ml double cream
30g smoked bacon lardons
10g butter

METHOD

1. Sweat onions and garlic in butter until soft; cut the cabbage into thin strips and cook in boiling salted water. Refresh in iced water, and strain.
2. Add cream to onion mix and reduce by half; add the cooked cabbage and smoked bacon lardons and continue to reduce until thick. Reserve for use.

Wheelers of St James parsnip puree

INGREDIENTS

745g parsnips, peeled and chopped
750ml milk
Salt

TO FINISH:

125ml cooking liquor
12g butter
5ml lemon juice
Cayenne pepper to season

METHOD

1. Peel and cut parsnip into pieces of same size.
2. Add to pan with milk, bring to boil, change pan to stop milk burning.
3. Add salt and cook gently until soft and tender; drain, keep milk.
4. Place in liquidiser and blitz until smooth, whilst adding the cooking milk and butter.
5. Pass through chinoise into another pan; place on a low heat, adding salt, pepper and lemon juice.
6. Season to taste.
7. Store in fridge with a cartouche.

Wheelers of St James venison marmalade

INGREDIENTS

2 bottles of red wine
400ml ruby port
750g onions
15g black peppercorns, crushed
2 bay leaves
3g thyme sprig
5g juniper berries, crushed
6 cloves

METHOD

1. Caramelise the onions until light golden and set aside.
2. Reduce red wine by three-quarters, adding the remaining ingredients.
3. Pour reduced liquid over caramelised onions.
4. Marinade for 24 hours before serving.

APPENDIX I: A Selection of Strong Trophies

R. Fraser 1993; gold,140 CIC.

H. Schreiber 1987, gold.

H-G. Oppermann 1989; gold, 131 CIC.

J. Klinke 1989; gold, 134 CIC.

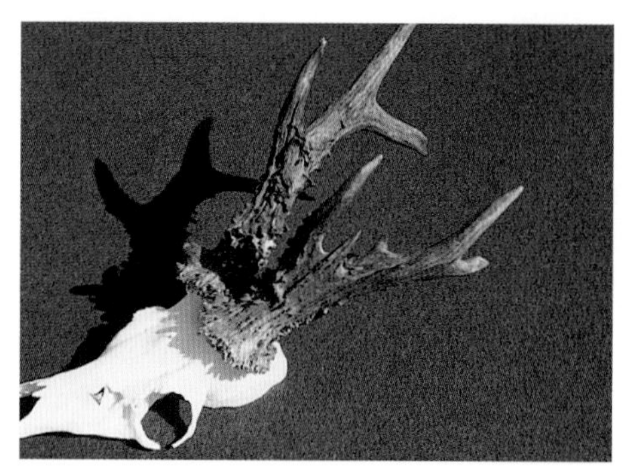

J. Schmit 1990; gold, 136CIC.

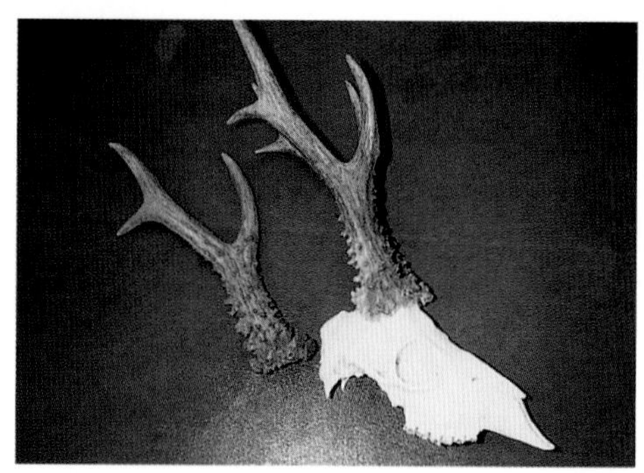

J. Lopez Bravo 1992; gold, 144 CIC.

W. Murlins 1993; gold, 144 CIC.

J. Grun 1994; gold 132 CIC.

J. Schmit 1994; gold, 135 CIC.

J. Poppen 1991; gold 130 CIC.

Furst Löwenstein 1996; gold 147 CIC.

F. Haschenmermes 1996; gold 138 CIC.

I. Boterman 1989; gold 150 CIC.

R. Siegmund 2000; gold, 147 CIC.

Furst Löwenstein 2003; gold 132 CIC.

R. Rashidian 2003, 134 CIC.

P. Vanhauter 2002; gold 131 CIC.

J. Grun 2002; gold, 133 CIC.

Road traffic accident.

C. Laval 1998; silver.

Furst Löwenstein 1997; silver.

Road traffic accident 1991; silver.

Road traffic accident.

H. Norman 1998; silver.

APPENDIX II:
A Selection of Bizarre Trophies

G. Kohn. *H. Norman.* *J-M. Grun.*

J. Poppen. *H-G. Oppermann.*

Prince D. Löwenstein.

E. Duponselle.

Count Erbach.

L. Beckman.

J. Grun.

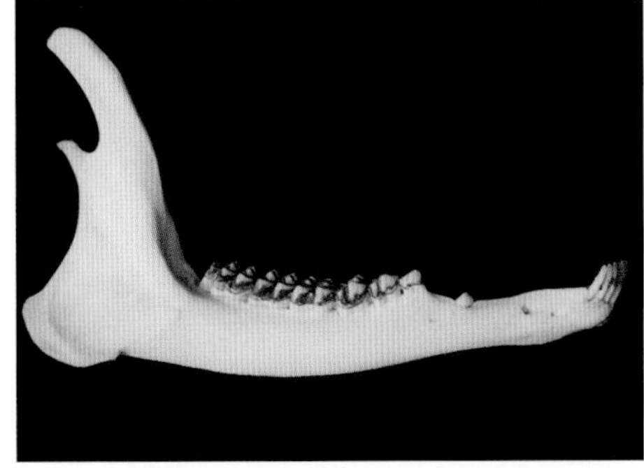

Roe with lower canines.

Roe with upper canines.

White-faced doe.

G. Boellertz.

J. Poppen, gold.

J.Poppen, gold.

J. Poppen, coalesced.

J. Schmit.

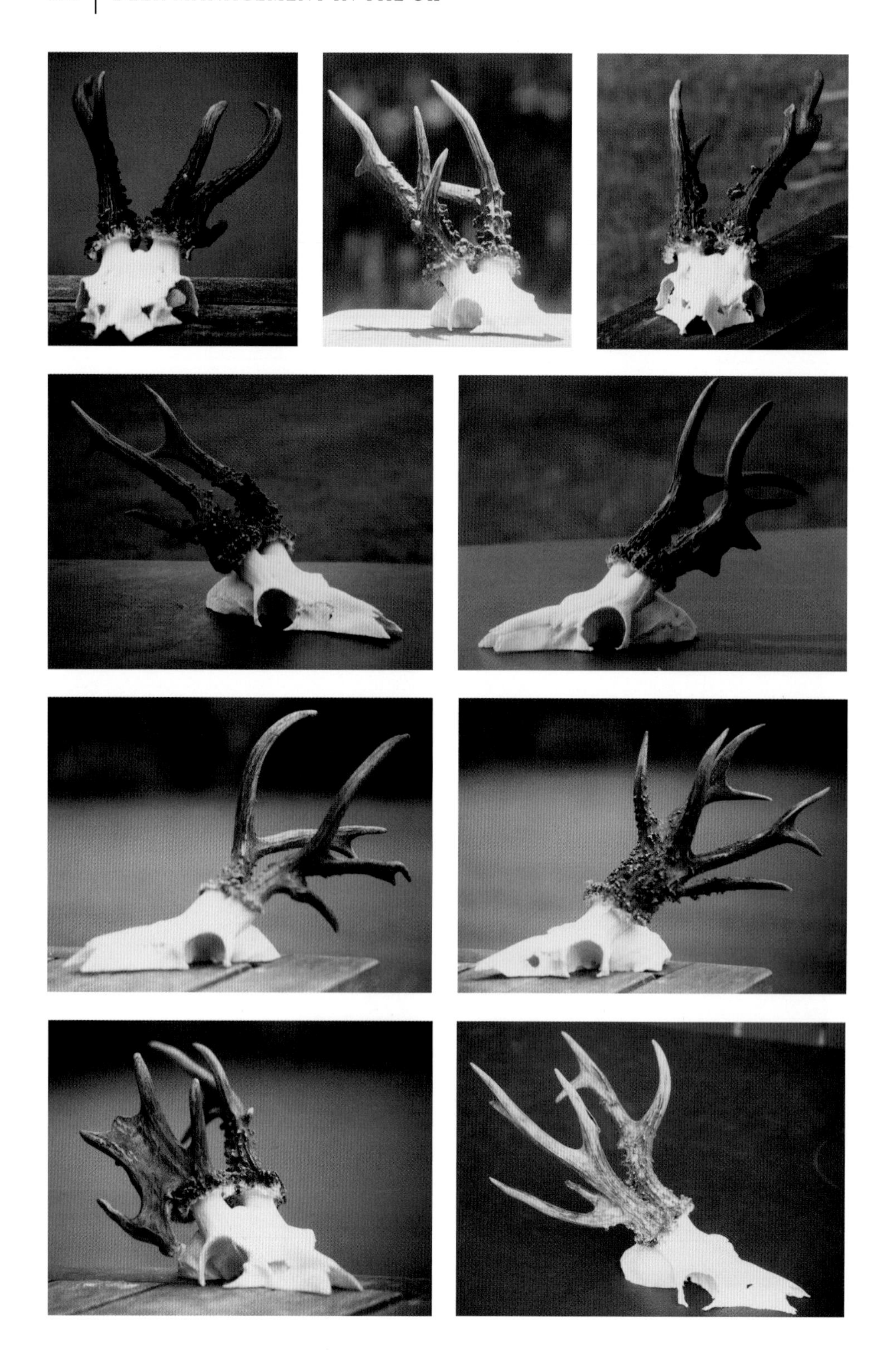